Copyright ⓒ 2010 Korea National Arboretum

Published by GeoBook Publishing Co.
  Officetel Rm 1321, 34, Sajik-ro 8-gil, Jongno-gu, Seoul, 110-872, KOREA
  Tel:+82-2-732-0337, Fax:+82-2-732-9337, e-mail:geo@geobook.co.kr

All rights reserved. No part of this book may be reproduced, stored in a retrieval system, or transmitted in any form or by any means, electronic, mechanical, photocopying, recording, or otherwise, without written permission from the copyright owner.

ISBN 978-89-94242-00-2 03600

Printed in Korea

세밀화로 보는
약용식물

Botanical Art of Korean Medicinal Plants

국립수목원 지음

GEOBOOK 지오북

책을 펴내며...

전 세계적으로 생물자원의 중요성이 강조되고 있는 이 시대에 우리와 함께 삶의 공간을 구성하는 식물 역시 매우 중요한 한 부분을 차지하고 있다는 것을 부인할 수 없을 것입니다. 오늘날 이러한 관심은 자생식물 분야까지 영역을 넓혀 다양한 방법으로 표현되고 있는 것 같습니다.

국립수목원에서는 2003년부터 살아있는 식물체의 영양형질인 잎, 뿌리, 줄기와 생식형질인 꽃의 미세구조, 열매, 종자 등을 한 장의 그림으로 담아내는 식물세밀화를 제작하고 있습니다. 식물세밀화는 식물연구의 중요한 산물로서 식물의 형태학적인 구조와 특징을 정확하고 섬세하게 표현해내는 고도의 전문성과 식물이 가진 고유한 아름다움을 돋보이게 하는 예술성을 겸비한 작품입니다.

이러한 식물연구자와 세밀화가의 오랜 노력이 만들어낸 결과물을 모아서 2005년에는 500년의 역사를 가진 광릉 숲에 자생하는 식물을 주제로 묶어 『세밀화로 보는 광릉 숲의 풀과 나무』를 발간하여, 식물세밀화란 또 하나의 문화콘텐츠(Culture Content)를 일반인들에게 널리 소개한 바 있습니다.

또한 매년 특정 주제를 바탕으로 '식물 세밀화 순회 전시회'를 개최하고 있습니다.

2008년에는 그 동안 제작된 식물세밀화 300여 점 중에서 약용식물로 질환을 치료하는 데 있어 각기 효능을 나타내는 식물 45점을 선별하여 '세밀화로 만나는 약용식물' 작품 전시회를 열고 약용식물에 대한 관심과 자원으로서의 중요성을 알리는 계기를 마련하였습니다.

이번에 발간하는 『세밀화로 보는 약용식물』은 금년에 전시한 작품을 포함해서 일반적으로 널리 이용되는 약용식물 세밀화 76점을 수록하였습니다. 이 책을 접함으로써 여러분들은 무엇보다 식물에 대한 아름다움을 느끼고, 식물에 대한 이해의 폭을 넓히며 식물에 대한 관심은 물론 새로운 가치를 인식할 수 있는 좋은 기회가 될 것으로 생각합니다.

끝으로 작품 활동의 여건이 좋지 않은 환경에서도 식물을 꾸준히 만나고 표현해서 식물세밀화를 완성하는 세밀화가 여러분들께 감사와 격려의 마음을 전합니다.

2009년 12월
국립수목원장

Prologue

The importance of biological resources is emphasized more than ever. It's indisputable that plants are more valuable than any other living organisms since our lives completely depend on them in many aspects. The interest in biological resources has translated into studies of wild plants and progressed in various ways.

Since 2003, Korea National Arboretum has embarked on the publication of botanical illustrations in which both the vegetative traits such as leaves, roots and stalks and the reproductive traits such as flowers, fruits and seeds are drawn in detail. The botanical illustrations are not only artistic master pieces enhancing the plant's unique characteristic beauty but also validates the results of the botanical research by delicately describing the accurate morphological structure and character of the plants.

As a result of endless efforts and dedications by botanists and artists, Korea National Arboretum published "Illustrated Herbs and Trees from Kwangreung Forest" in 2005, covering numerous native plants conserved in the 500-year-old forest. Since the botanical illustration had not been well aware until then, the book contributed to introduce a new genre of botanical arts as a cultural content to the public.

In addition, Korea National Arboretum has held the botanical arts exhibition tour with a specific theme every year. Especially, the 2008 exhibition entitled "Illustrated Medicinal

Plants" provided the illustrations of 45 medicinal plants selected from 300 ones, by which we were able to call the public attention to the medicinal plants and their importance as natural resources.

This book, 「The Botanical Art of Korean Medicinal Plants」, contains 76 botanical illustrations of important medicinal plants, including the ones presented in the 2008 exhibition. Encountering this book would help the public feel the beauty of plants itself as well as broaden the understanding of their importance. Also, it would provide a golden opportunity for the public to recognize plants' novel values.

My deepest gratitude is extended to the botanical artists for working in challenging environment. Your craft is greatly admired by the readers, making possible the close-knit marriage of text and image. The admiration, of course, is balanced with utter jealousy of your working conditions!

December, 2009
The Direct General of Korea National Arboretum

발간을 축하하며...

국립수목원에서 약초에 관련된 세밀화 그림을 모은 『세밀화로 보는 약용식물』이 발간됨을 축하드립니다.
국립수목원에서는 약 6년 전부터 식물세밀화의 중요성을 인식하고 세밀화 제작에 심혈을 기울이고 있습니다. 이 책은 그 중에서 약재로 유용한 식물 그림들을 모아놓은 책이란 점에서 매우 뜻 깊은 일이라 생각됩니다.

식물세밀화(Botanical Art & Illustration)란 무엇일까요? 단지 식물을 자세히 그리는 것일까요? 사람들은 그렇게 생각할 수 있습니다. 그러나 그것을 그리는 작가의 마음과 눈과 손은 그렇지 않습니다. 거기에는 세밀한 관찰과 정확한 표현과 올바른 색감과 식물학적 정확성과 더불어 화면의 창의성을 부여해 아름다움을 나타내는 복합적인 미적 감각을 지녀야 하고 작가들은 그것을 고민하며 작품에 임하게 됩니다. 때로는 운 좋게 자연의 모습 자체가 가감이 필요 없는 아름다운 형태나 선을 가지고 있다면 작가로서는 행운을 만난 것이며, 그렇지 않은 경우에는 많은 고심을 통해 아름답게 표현되었을 때 또한 희열을 느낄 수 있습니다. 그리고 그것은 보는 이로 하여금 공감이란 선물로 제공되는 것이지요.

본래 세밀화 역사를 보면 그 시작은 인간에게 유익한 것과 위험한 것을 구별하기 위한 그림이었습니다. 초창기에는 단지 약용이나 식물학적 목적으로 그려지던 세밀화가 세월이 흘러 많은 작가가 배출되면서 작가들의 욕구는 더 높은 곳을 지향하였습니다. 그들은 식물학연구 분야에는 없어서는 안 될 중요한 역할을 감당하면서 아름다움도 함께 추구하는 작업을 하였습니다.

특히 약용식물 그림을 이야기할 때 빠질 수 없는 역사적 인물이 있습니다. 식물학자이며 의사였던 레온하르트 푹스(Leonhart Fuchs, 1501~1566)는 약용식물을 그림으로 제작하고 1542년 『De Historia Stirpium』이라는 책으로 출판하여 대중에게 약초에 대한 지식을 바르게 알리는 데 큰 기여를 하였습니다.

국립수목원도 이와 같은 일을 하기 위하여 어려운 여건 가운데에서 지속적인 작업을 하는 것으로 알고 있습니다. 이 일은 한반도의 식물에 대한 기초자료를 만드는 것으로서 매우 중요한 일이라 생각됩니다. 이 일에 참여한 작가들의 노력과 열정이 약용식물 세밀화를 일반인에게 이해시키는 데 큰 역할을 하리라 생각됩니다. 바라기는 앞으로 우리나라에도 훌륭한 보태니컬 아티스트들이 많이 배출되어 보다 우수하고 품격 높은 식물세밀화들이 나와서 대중들의 사랑을 듬뿍 받는 때가 오기를 기대해 봅니다.

그리고 이만큼이라도 약용식물 세밀화의 올바른 모습이 나타나도록 노력하고 책으로 엮어 세상에 드러나게 해주신 국립수목원과 산림청의 관계자들에게 감사의 말을 전합니다.

2009년 12월
사단법인 한국식물세밀화협회 회장 구 지 연

Celebrating the Publication...

I deeply celebrate that Korea National Arboretum publishes 「The Botanical Art of Medicinal Plants」, the book collecting illustrations on medicinal plants. As we all know, Korea National Arboretum realized the importance of botanical arts and illustrations 6 years ago and has devoted to producing them. It is believed that this book is one of the valuable outcomes from this endeavor in that the main purpose is to illustrate medicinal plants deliberately.

What is the botanical art and illustration? Is it simply drawing plants in detail? People may think so. However, the heart and hands of painters are different. They should have compound aesthetic senses not only for the botanical accuracy obtained from close observation, the exact expression and the proper color sense, but also for the beautiful description granted by originality. Considering these prerequisites, painters ought to fully engage themselves in works. It is a fortune for them to encounter natural objects of beautiful shapes and lines which do not require any modification at all. In other cases, they strive to express objects exactly and beautifully, which gives birth to rapture when a satisfactory result is made. Furthermore, the artworks present viewers with happiness from and communion with nature.

The history of botanical art and illustration started to show the distinction between beneficial and perilous species. This art which was aimed for medical or botanical purposes in the beginning came to aim for higher artistic fulfillment as more talented artists participated. Not only did they play indispensable and important roles in research on botany, but also pursued artistic beauty and genuineness.

Especially, there is a historic person without whom we cannot mention botanical arts of medicinal plants. Leonhart Fuchs (1501~1566), the doctor and botanist, painted medicinal plants and then published the book, 「De Historia Stirpium」, in 1542 to make great contribution to informing the public of herb age.

I have well realized that Korea National Arboretum has conducted continuous works for completing the same mission. I believe that this task has a vital importance in the establishment of research foundation for plants in the Korea Peninsula. Passions and efforts of those who participated in this project would play a significant role in understanding the necessity and consequence of the botanical art of medicinal plants. I hope that as more outstanding botanical artists are produced in Korea, excellent and refined botanical illustrations painted by them will be showered with affection by people. Finally, I express my deepest gratitude to persons concerned in Korea National Arboretum and the Office of Forestry who compiled the book and exerted themselves for rightly depicting medicinal plants on canvases.

December, 2009
Korean Association of Botanical Arts and Illustrations
President Koo, Jee-Yeon

차 례

책을 펴내며..............004
발간을 축하하며........008

고란초과
고란초...................016
산일엽초.................018
우단일엽.................020

삼백초과
삼백초...................022

꼬리겨우살이과
겨우살이.................024

쥐방울덩굴과
등칡.....................026
쥐방울덩굴...............028

석죽과
패랭이꽃.................030

미나리아재비과
꿩의바람꽃...............032

으름덩굴과
으름덩굴.................034

매자나무과
삼지구엽초...............036
깽깽이풀.................038

양귀비과
애기똥풀.................040

현호색과
점현호색.................042
현호색...................044

장미과
뱀딸기...................046

괭이밥과
큰괭이밥.................048

다래나무과
다래.....................050

제비꽃과
제비꽃...................052

두릅나무과
가시오갈피...............054
인삼.....................056

산형과
섬시호...................058

층층나무과
산수유...................060

앵초과
참좁쌀풀.................062
큰까치수염...............064

용담과
용담.....................066
과남풀...................068
큰구슬붕이...............070

박주가리과
박주가리.................072

마편초과
누리장나무...............074

꿀풀과
꽃향유....................076
벌깨덩굴..................078
백리향....................080
섬백리향..................082

꼭두서니과
선갈퀴....................084

인동과
인동......................086

초롱꽃과
도라지....................088

국화과
벌개미취..................090
삽주......................092
큰엉겅퀴..................094
바늘엉겅퀴................096
산국......................098
좀씀바귀..................100
왕씀배....................102
미역취....................104

흑삼릉과
흑삼릉....................106

천남성과
둥근잎천남성..............108
천남성....................110
두루미천남성..............112
점박이천남성..............114
큰천남성..................116
반하......................118

닭의장풀과
닭의장풀..................120

백합과
달래......................122
두메부추..................124
산마늘....................126
산부추....................128
울릉산마늘................130
참산부추..................132
한라부추..................134
윤판나물..................136
큰애기나리................138
말나리....................140
섬말나리..................142
참나리....................144
삿갓나물..................146
각시둥굴레................148
왕둥굴레..................150
층층둥굴레................152
금강애기나리..............154
연영초....................156
산자고....................158

난초과
자란......................160
약난초....................162
석곡......................164
천마......................166

용어해설..................168
찾아보기..................172
참고문헌..................176

Contents

Prologue 006
Celebrating the
            Publication......010

Polypodiaceae
Crypsinus hastatus 016
Lepisorus ussuriensis 018
Pyrrosia linearifolia 020

Saururaceae
Saururus chinensis 022

Loranthaceae
Viscum album var. *coloratum*
....024

Aristolochiaceae
Aristolochia manshuriensis
....026
Aristolochia contorta 028

Caryophyllaceae
Dianthus chinensis 030

Ranunculaceae
Anemone raddeana 032

Lardizabalaceae
Akebia quinata 034

Berberidaceae
Epimedium koreanum 036
Jeffersonia dubia 038

Papaveraceae
Chelidonium majus var.
       asiaticum........................040

Fumariaceae
Corydalis maculata 042
Corydalis remota 044

Rosaceae
Duchesnea indica 046

Oxalidaceae
Oxalis obtriangulata 048

Actinidiaceae
Actinidia arguta 050

Violaceae
Viola mandshurica 052

Araliaceae
Eleutherococcus senticosus
....054
Panax ginseng 056

Apiaceae
Bupleurum latissimum 058

Cornaceae
Cornus officinalis 060

Primulaceae
Lysimachia coreana 062
Lysimachia clethroides 064

Gentianaceae
Gentiana scabra 066
Gentiana triflora var.
    japonica 068
Gentiana zollingeri 070

Asclepiadaceae
Metaplexis japonica 072

Verbenaceae
Clerodendrum trichotomum..074

Lamiaceae
Elsholtzia splendens 076
Meehania urticifolia 078
Thymus quinquecostatus ..080
Thymus quinquecostatus var. *japonica* 082

Rubiaceae
Asperula odorata 084

Caprifoliaceae
Lonicera japonica 086

Campanulaceae
Platycodon grandiflorum .088

Asteraceae
Aster koraiensis 090
Atractylodes ovata 092
Cirsium pendulum 094
Cirsium rhinoceros 096
Dendranthema boreale 098
Ixeris stolonifera 100
Prenanthes ochroleuca 102
Solidago virgaurea subsp. *asiatica* 104

Sparganiaceae
Sparganium erectum 106

Araceae
Arisaema amurense 108
Arisaema amurense for. *serratum* 110
Arisaema heterophyllum ..112
Arisaema peninsulae 114
Arisaema ringens 116
Pinellia ternata 118

Commelinaceae
Commelina communis 120

Liliaceae
Allium monanthum 122
Allium senescens 124
Allium microdictyon 126
Allium thunbergii 128
Allium ochotense 130
Allium sacculiferum 132
Allium taquetii 134
Disporum uniflorum 136
Disporum viridescens 138
Lilium distichum 140
Lilium hansonii 142
Lilium lancifolium 144
Paris verticillata 146
Polygonatum humile 148
Polygonatum robustum 150
Polygonatum stenophyllum 152
Streptopus ovalis 154
Trillium kamtschaticum 156
Tulipa edulis 158

Orchidaceae
Bletilla striata 160
Cremastra variabilis 162
Dendrobium moniliforme..164
Gastrodia elata..................166

Glossary 168
Index....................................172
References...........................176

고 란 초 *Crypsinus hastatus* (Thunb.) Copel.
고 란 초 과 | Polypodiaceae

| 식 물 | 전국 산지의 바위나 고목에 착생하여 자라는 상록성 여러해살이풀로 드물게 관찰된다. 뿌리줄기는 옆으로 길게 벋으며 잎이 드문드문 달린다. 잎은 1장씩 달리며 타원상 피침형이거나 드물게 가장자리가 2~3갈래로 갈라진다. 포자낭군은 원형으로 갈색이며 2줄로 달린다.
| 분 포 | 한국, 일본, 대만, 중국
| 약 재 명 | 아장금성초(鵝掌金星草)
| 효 능 | 종기, 종창, 악창

| Description | Evergreen perennial, on shallow soil on rocks or dead trees, widely distributed in mountains but rarely observed. Leaves sparsely born along the slender creeping rhizomes, simple, ovate-lanceolate or sometimes 2 to 3-lobed. Sori orbicular, brown, born in two rows along the veins.
| Distribution | Korea, Japan, Taiwan, China
| Medicinal Name | A-jang-geum-seong-cho(鵝掌金星草)
| Effect | Boil, swelling, obstinate abscess

| 이용부위 Parts Used |
지상부 Above-ground parts

산일엽초 *Lepisorus ussuriensis* (Regel & Maack) Ching
고 란 초 과 | Polypodiaceae

식 물	전국 산지의 바위나 고목에 착생하여 자라는 상록성 여러해살이풀이나. 뿌리줄기는 가늘고 길게 옆으로 벋고 잎이 드문드문 달린다. 잎은 1장씩 달리고 선상 피침형이다. 포자낭군은 원형으로 잎의 윗부분에 2줄로 달린다.
분 포	한국, 일본, 만주, 우수리, 사할린
약 재 명	오소리와위(烏蘇里瓦韋)
효 능	요로감염증, 신우신염, 기침, 백일해, 소변출혈, 각혈, 인후염, 구강염

Description	Evergreen perennial, widely distributed on shallow soil on rocks or dead trees in mountains, but rarely observed. Leaves sparsely born along the slender creeping rhizomes, simple, linear-lanceolate. Sori orbicular, brown, born in two rows beside the midribs.
Distribution	Korea, Japan, Manchuria, Ussuri, Sakhalin
Medicinal Name	O-so-ri-wa-wi(烏蘇里瓦韋)
Effect	Urinary tract infection, pyelonephritis, cough, whooping cough, hematuria, hemoptysis, sore throat, stomatitis

| 이용부위 Parts Used |
지상부, 뿌리줄기
Above-ground parts and root stock

우단일엽 *Pyrrosia linearifolia* (Hk.) Ching
고 란 초 과 | Polypodiaceae

| 식 물 | 전국 산지의 바위나 고목에 착생하여 자라는 상록성 여러해살이풀로 드물게 분포한다. 뿌리줄기는 길게 옆으로 뻗으며 인편이 밀생하고 잎이 드문드문 달린다. 잎은 1장씩 달리고 선형이며 양면에 다갈색의 별 모양 털이 밀생한다. 포자낭군은 타원형 또는 원형으로 잎의 윗부분 주맥 양쪽에 2줄로 배열된다.

| 분 포 | 한국, 일본, 대만, 중국, 만주

| 약 재 명 | 소석위(小石葦)

| 효 능 | 발열성 소아 경련발작, 외상출혈

| Description | Evergreen perennial, on shallow soil on rocks or dead trees, widely distributed in mountains but rarely observed. Leaves sparsely born along the scaled rhizomes, simple, linear, both sides covered with star-shaped hairs. Sori oblong or orbicular, arranged in two rows beside midribs.

| Distribution | Korea, Japan, Taiwan, China, Manchuria

| Medicinal Name | So-seog-wi(小石葦)

| Effect | Febrile infant convulsive seizure, open injury and bleeding

| 이용부위 **Parts Used** |
지상부 Above-ground parts

삼백초

Saururus chinensis (Lour.) Baill.
삼백초과 | Saururaceae
Chinese Lizard's Tail

식 물	제주도의 해안가 습지에 드물게 자라는 여러해살이풀이다. 뿌리줄기는 백색으로 길게 옆으로 벋는다. 줄기는 곧추서고 높이 50~100cm 정도이다. 잎은 어긋나고 넓은 난형으로 표면은 연한 녹색, 뒷면은 연한 백색이지만 윗부분의 잎 2~3장은 표면이 백색으로 변한다. 꽃은 6~8월에 피고 백색으로 총상화서에 달린다. 열매는 삭과로 둥글고 8~9월에 익는다.
분 포	한국, 일본, 중국, 필리핀
약재명	삼백초(三白草)
성 분	methyl-n-nonyl-ketone, tannin, quercitrin, isoquercitrin, avicularin
효 능	황달, 종기, 악창, 백대하

Description	Perennial herb, rare in wet lands along the Jeju Island seashore. Commonly cultivated as medicinal plant. Roots white, rhizomatous. Stem erect, 50-100cm high. Leaves alternate, broadly ovate, the upper surface pale green, the lower surface pale white, 2-3 leaves at the top of the stem white. Flowers white, born in a raceme, blooming in Jun.-Aug. Fruits capsule, globular, ripening in Aug.-Sep.
Distribution	Korea, Japan, China, Philippines
Medicinal Name	Sam-back-cho(三白草)
Chemical components	Methyl-n-nonyl-ketone, tannin, quercitrin, isoquercitrin, avicularin
Effect	Jaundice, boil, obstinate abscess, leukorrhoea

| 이용부위 Parts Used |
지상부 Above-ground parts

겨우살이

Viscum album var. *coloratum* (Kom.) Ohwi
꼬리겨우살이과 | Loranthaceae
Mistletoe

| 식　물 | 참나무류, 밤나무, 자작나무 등과 같은 활엽수에 기생하는 상록성 관목으로 둥지같이 모여 자라고 지름이 1m에 달하는 것도 있다. 황록색인 가지는 매끈하고 Y자로 갈라진다. 옅은 황색 꽃은 2~3월에 피고 가지 끝에 보통 3개씩 달린다. 열매는 장과로 10~12월에 반투명한 연한 노란색으로 익는다. |

| 분　포 | 한국, 일본, 대만, 중국, 유럽, 아프리카 |

| 약재명 | 곡기생(槲寄生) |

| 성　분 | oleanolic acid, β-amyrin, mesoinositol, flavonoid 화합물 |

| 효　능 | 동통, 무력증 |

Description Evergreen shrub, parasitic on branches of broad leaf trees such as oak, chestnut tree, white birch, etc. widely distributed in mountains. Stems born in a cluster like a nest, up to 1m in diameter, yellowish green, Y-shaped branching. Flowers light yellow, three flowers born at the branch tip, blooming in Feb.-Mar. Fruits berry, translucent light yellow, ripening in Oct.-Dec.

Distribution Korea, Japan, Taiwan, China, Europe, Africa

Medicinal Name Gok-gi-saeng(槲寄生)

Chemical components Oleanolic acid, β-amyrin, mesoinositol, flavonoid compounds

Effect Pain, asthenia

| 이용부위 **Parts Used** |
전초 All parts

Seung-Hyun Yi 2005

등칡

Aristolochia manshuriensis Kom.
쥐방울덩굴과 | Aristolochiaceae
Manchurian Birthwort, Manchurian Dutchman's pipe

| 식 물 | 깊은 산의 기슭 또는 계곡에서 드물게 자라는 여러해살이 낙엽성 덩굴식물이다. 새 가지는 녹색이지만 2년지는 회갈색이다. 잎은 넓은 심장형으로 거치가 없다. 색소폰 모양(U자)으로 구부러진 황색꽃은 5월경에 엽액에서 한 개씩 나온다. 꽃의 겉은 연녹색, 안쪽의 중앙부는 연갈색, 밑부분은 자흑색, 윗부분은 자갈색으로 반점이 있다. 열매는 삭과로 10~11월경에 익는다.
| 분 포 | 한국, 중국 동북부, 우수리
| 약 재 명 | 관목통(關木通)
| 성 분 | aristolochic acid, oleanolic acid, hedragenin
| 효 능 | 방광염, 유즙분비, 구내염

| Description | Perennial vine, deciduous, rare in foothills of deep mountains or valleys. New branches green, 2-year branches gray-brown. Leaves broadly cordate, entire. Flowers yellow at the mouth part, light green on the outer tube, light brown on the center of the inner part, purplish black on the lower part, and purplish brown and dotted on the upper part, bending like a saxophone, solitary, born at the axil, blooming in May. Fruits capsule, oblong, ripening in Oct.-Nov.
| Distribution | Korea, Northeast China, Ussuri
| Medicinal Name | Gwan-mok-tong(關木通)
| Chemical components | Aristolochic acid, oleanolic acid, hedragenin
| Effect | Cystitis, lactation, stomatitis

| 이용부위 Parts Used |
줄기 Stem

쥐방울덩굴

Aristolochia contorta Bunge
쥐방울덩굴과 | Aristolochiaceae
Northern Dutchmanspipe

| 식 물 | 제주도를 제외한 전국의 산과 들이나 숲 가장자리에 자라는 여러해살이 덩굴식물이다. 잎은 어긋나며 심장형 또는 넓은 심장형으로 흰빛이 도는 녹색이다. 꽃은 7~8월에 피며 엽액에서 1개씩 달린다. 꽃받침은 통처럼 밑부분이 둥글게 커지고 윗부분이 좁아졌다가 나팔처럼 벌어지며 한쪽 열편이 길게 뾰족해진다. 열매는 삭과로서 큰 구형이고 6개의 골이 있다. 성숙하면 6개로 갈라진 다음 6개로 갈라지는 열매자루에 매달려서 낙하산같이 된다.

| 분 포 | 한국, 일본, 만주

| 약 재 명 | 마두령(馬兜鈴)-열매, 청목향(靑木香)-뿌리

| 성 분 | 마두령-aristolochic acid, alkaloid
청목향-aristolone, aristolovhic acid, allantoin, debilic acid

| 효 능 | 마두령-해수, 가래, 천식, 치질, 원발성고혈압
청목향-흉복부 팽만, 장염, 이질, 종기

| Description | Perennial vine, deciduous, widely distributed (except the Jeju Island) but rarely observed in open fields along the edge of forests. Leaves alternate, cordate or broadly cordate, white-tinged green. Flowers born in axils, blooming in Jul.-Aug. Calyx tube enlarged at the bottom, bending like a saxophone, widely opened at the mouth part, a lobe attenuated in one side. Fruits capsule, globose, six-grooved, dehisced fruits similar to a parachute with 6 strings divided from a pedicel.

| Distribution | Korea, Japan, Manchuria

| Medicinal Name | Ma-du-ryeong(馬兜鈴) - fruit, cheong-mok-hyang(靑木香) - root

| Chemical components | Ma-du-ryeong - aristolochic acid, alkaloid; cheong-mok-hyang - aristolone, aristolovhic acid, allantoin, debilic acid

| Effect | Ma-du-ryeong - cough, sputum, asthma, hemorrhoids, essential hypertension; cheong-mok-hyang - abdominal fullness, enteritis, dysentery, boil

| 이용부위 Parts Used |
열매, 뿌리 Fruit and root

패랭이꽃

Dianthus chinensis L.
석죽과 | Caryophyllaceae
Pink, Chinese Pink

식 물	전국 각지의 낮은 지대, 건조한 곳 또는 냇가 모래땅에 자라는 여러해살이풀이다. 줄기는 모여 나며 높이 30cm 내외로서 위에서 가지가 갈라진다. 잎은 마주나고 밑부분에서 합쳐져서 원줄기를 둘러싸며 잎의 가장자리가 밋밋하다. 붉은색의 꽃은 6~8월경에 가지 끝에 한 개씩 피고 꽃잎은 5개로 가장자리가 얕게 갈라지며 바로 그 밑에 짙은 무늬와 더불어 긴 털이 약간 있다. 열매는 삭과로 9~10월경에 익으며 꽃받침으로 싸여 있다.
분 포	한국, 중국
약재명	구맥(瞿麥)
성 분	휘발성 정유, saponin
효 능	방광염, 요도염, 급성신우신염, 월경폐색, 종기

Description	Perennial herb, widely distributed in lowlands, arid places, or sandy stream banks. Stems erect, clustered, about 30cm tall, branching above. Leaves opposite, lanceolate, entire; the bases of opposite leaves enclosing the stem. Flowers red, solitary, born at the branch tip, blooming in Jun.-Aug.; petals five, shallowly serrate at the tip, sparsely long-haired below. Fruits capsule, oblong, enclosed by calyx tube, ripening in Sep.-Oct.
Distribution	Korea, China
Medicinal Name	Gu-maek(瞿麥)
Chemical components	Volatile essential oil, saponin
Effect	Cystitis, urethritis, acute pyelonephritis, dysmenorrhea, boil

| 이용부위 Parts Used |
지상부 Above-ground parts

꿩의바람꽃 *Anemone raddeana* Regel
미나리아재비과 | Ranunculaceae
Radde Anemone

| 식 물 | 전국 산지의 숲 속에 자라는 여러해살이풀이다. 육질인 뿌리줄기는 옆으로 벋고 굵다. 잎은 3장의 소엽이 3개씩 달리며 소엽은 끝이 3갈래로 깊게 갈라진다. 꽃은 4~5월에 피고 흰색으로 약간 자줏빛이 돌며 꽃줄기 위에 1개씩 달린다. 열매는 수과로 7월에 익는다.
| 분 포 | 한국, 일본, 만주, 아무르, 우수리
| 약재명 | 죽절향부(竹節香附)
| 효 능 | 사지마비, 요통, 종기, 외상

| Description | Perennial herb, widely distributed in the forests of mountains or mountain ridges. Root stocks fleshy, thick, rhizomatous. Leaves 2-3 trifoliate. Flowers white slightly tinged with purple, solitary, born at the scape tip, blooming in Apr.-May. Fruits achene, ripening in Jul.
| Distribution | KOREA, Japan, Manchuria, Amur, Ussuri
| Medicinal Name | Juk-jeol-hyang-bu(竹節香附)
| Effect | Quadriplegia, lumbago, boil, open injury

| 이용부위 Parts Used |
뿌리줄기 Root stock

으름덩굴

Akebia quinata (Thunb.) Decne.
으름덩굴과 | Lardizabalaceae
Five-Leaf Akebia, Chocolate Vine

| 식 물 | 황해도 이남에 나며 산과 들의 다소 습한 곳에 자라는 낙엽성 덩굴식물이다. 가지는 털이 없고 갈색이다. 5~6개의 소엽이 모인 손바닥 모양의 장상엽은 묵은 가지에서는 무리지어 나고 새 가지에서는 어긋난다. 소엽은 난형이거나 타원형이며 가장자리가 밋밋하고 끝이 약간 오목하다. 자줏빛을 띤 꽃은 4~5월경에 피고 엽액에서 총상으로 달린다. 수꽃은 작고 많이 달리며 암꽃은 크고 적게 달린다. 과피가 두터운 열매는 긴 타원형이며 10월에 자갈색으로 익는다.

| 분 포 | 한국, 일본, 중국
| 약 재 명 | 목통(木通)-줄기, 예지자(預知子)-열매
| 성 분 | akebin, hederagenin, oleanolic acid
| 효 능 | 신우신염, 방광염, 요도염, 마비동통, 유즙분비

| Description | Deciduous vine, widely distributed in somewhat wet places of open fields and mountains south of Hwanghae-do. Stems glabrous, brown. Leaves 5-6 palmate, fasiculate at the old branches or alternate at the new branches, leaflets ovate or elliptical, entire, emarginate at the tip. Flowers purple, born in a raceme, blooming in Apr.-May male flowers small and many, and female flowers large and a few. Fruits follicle, purplish brown, oblong, ripening in Oct.; endocarp fleshy and pericarp thickened.
| Distribution | Korea, Japan, China
| Medicinal Name | Mokt-ong(木通) - stem, ye-ji-ja(預知子) - fruit
| Chemical components | Akebin, hederagenin, oleanolic acid
| Effect | Pyelonephritis, cystitis, urethritis, paralysis and pain, lactation

| 이용부위 Parts Used |
줄기, 열매 Stem and fruit

삼지구엽초

Epimedium koreanum Nakai
매자나무과 | Berberidaceae
Korean Epimedium

| 식 물 | 중부 이북 산지의 숲 속에서 드물게 자라는 여러해살이풀이다. 뿌리줄기는 옆으로 벋고 잔뿌리가 많다. 줄기는 30cm 정도까지 자라고 기부는 비늘 모양의 잎으로 둘러싸인다. 줄기 윗부분은 3개의 가지가 갈라지고 가지 끝마다 3개의 잎이 달려 삼지구엽초라 한다. 난형의 잎은 끝이 뾰족하고 밑부분은 심장형이며 잎 가장자리에 털 같은 잔톱니가 있다. 미색의 꽃은 5~6월에 피고 줄기 끝에 총상으로 달려 밑을 향한다. 양끝이 뾰족한 원기둥 모양의 열매는 골돌로 7~8월경에 결실한다.
주 의_ 형태가 유사한 산꿩의다리, 참꿩의다리, 노루오줌 등을 대용품으로 사용하기도 하나 독성이 있고 효능이 다르므로 주의가 필요하다. |
분 포	한국, 중국 동북부
약 재 명	음양곽(淫羊藿)
성 분	flavonoid, 휘발성 정유, cholesterol, alkaloid
효 능	발기부전, 사지냉증, 피부마비, 구안와사, 반신불수

Description	Perennial herb, uncommon in the mountainous forests north of the central peninsula. Commonly cultivated as medicinal plant. Roots fibrous or rhizomatous. Stem up to 30cm tall, the base enclosed with scaly leaves, three-branched above, three leaves on the end of each branch (The Korean name, Samjiguyeopcho, came from the leaves with three branches and nine leaflets). Leaves ovate, acuminate at the apex, cordate on the base, spiny-haired at the margin. Flowers yellowish white, born in a raceme at the top of stem, blooming in May-Jun. Fruits follicle, cylindrical with a pointed tip, ripening in Jul.-Aug.
Distribution	Korea, Northeast China
Medicinal Name	Eum-yang-gwak(淫羊藿)
Chemical components	Flavonoid, volatile essential oil, cholesterol, alkaloid
Effect	Impotence, cold hands and feet, cutaneous paralysis, Bell's palsy, hemiplegia

| 이용부위 Parts Used | 전초 All parts |

깽깽이풀

Jeffersonia dubia (Maxim.) Benth. & Hook.f. ex Baker & S.Moore
매자나무과 | Berberidaceae
Chinese Twinleaf

식 물	제주도를 제외한 전국의 산중턱 아래 골짜기에서 드물게 자라는 여러해살이풀이다. 뿌리줄기는 짧으며 많은 잔뿌리가 있다. 잎은 뿌리줄기 끝에서 모여 나고 긴 잎자루 끝에 1개씩 달리며 가장자리가 물결 모양이다. 꽃은 4~5월에 자줏빛을 띤 붉은색으로 피고 잎보다 먼저 밑동에서 나온 1~2개의 꽃줄기 끝에 1개씩 달린다. 열매는 삭과로 넓은 타원형이다.
분 포	한국, 만주, 아무르, 우수리
약재명	선황련(鮮黃連)
성 분	berberin
효 능	편도선염, 결막염, 소화불량, 식욕감퇴, 구토, 장염, 복통, 설사, 이질, 코피, 토혈

Description	Perennial herb, widely distributed (except the Jeju Island), rarely observed in mountain foothills. Root stocks short and thick with many slender roots. Leaves radical, clustered, leaflets long-petioled orbicular, wavy at the margin. Flowers purple-colored red, solitary at the scape tip, blooming before leaf emergence in Apr.-May. Fruits capsule, broadly oblong.
Distribution	Korea, Manchuria, Amur, Ussuri
Medicinal Name	Seon-hwang-nyeon(鮮黃連)
Chemical components	Berberin
Effect	Tonsillitis, conjunctivitis, dyspepsia, appetite decreased, vomiting, enteritis, abdominal pain, diarrhoea, dysentery, nosebleed, hematemesis

| 이용부위 Parts Used |
| 뿌리 Root |

2004. SEUNG-HYUN YI.

애기똥풀

Chelidonium majus var. *asiaticum* (Hara) Ohwi
양귀비과 | Papaveraceae
Asian Celandine, Asian Greater Celandine

식 물	전국 각지의 길가나 풀밭에서 자라는 여러해살이풀이다. 뿌리는 곧고 땅 속 깊이 들어가며 노란빛을 띤 주황색이다. 가지가 많이 갈라지는 줄기는 속이 비어 있고 높이 30~80cm 정도이며 흰색의 곱슬털이 빽빽하게 나고 상처를 내면 진노랑색 액이 나온다. 어긋나는 잎은 1~2회 새깃 모양으로 갈라지고 잎 뒷면은 흰색, 표면은 녹색이다. 황색의 꽃은 5~8월경에 원줄기와 가지 끝에 달린다. 열매는 삭과로 좁은 원기둥 모양이다.
분 포	한국, 일본, 중국 동북부, 사할린, 몽골, 시베리아, 캄차카
약재명	백굴채(白屈菜)
성 분	chelidonine, protopine, stylopine, allocryptopine, berberine
효 능	급만성위장염, 위·십이지장궤양, 복부동통, 이질, 황달형간염, 피부궤양, 결핵, 옴, 버짐

Description	Perennial herb, widely distributed in roadsides or grasslands. Taproots yellowish orange. Stems many-branched, hollow, 30-80cm tall, with densely curled white hairs, oozing yellow sap when cut. Leaves alternate, pinnately compound, whitish below and green above. Flowers yellow, born in an umbellate cyme, blooming in May-Aug. Fruits capsule, narrowly cylindrical.
Distribution	Korea, Japan, Northeast China, Sakhalin, Mongolia, Siberia, Kamchatka
Medicinal Name	Baek-gul-chae(白屈菜)
Chemical components	Chelidonine, protopine, stylopine, allocryptopine, berberine
Effect	Acute/chronic gastritis, gastric/duodenal ulcer, abdominal pain, dysentery, iteric hepatitis, skin ulcer, tuberculosis, scabies, trichophytia

| 이용부위 Parts Used |
지상부 Above-ground parts

점현호색

Corydalis maculata B.U.Oh & Y.S.Kim
현호색과 | Fumariaceae

| 식 물 | 깊은 산에 나는 여러해살이풀이다. 높이 20cm 정도로 식물체는 연약하다. 잎은 어긋나며 1~2회 갈라지고 흔히 흰색 반점이 있다. 청색인 꽃은 4~5월에 피고 3~18개가 원줄기 끝에 총상으로 달린다. 열매는 삭과이며 납작한 방추형이고 양끝이 좁으며 끝에 암술머리가 달려 있다. 종자는 거의 2열 배열하고 광택이 있다.

| 분 포 | 특산(강원, 경기, 경북, 충북)

| 약재명 | 현호색(玄胡索)

| 성 분 | corydaline, dl-tetrahydropalmatine, protopine, l-tetrahydrocoptisine

| 효 능 | 생리통, 복통, 요통, 두통, 신경통

| Description | Perennial herb, rare in deep mountains. Stems 20cm tall, procumbent. Leaves alternate, 2-3 pinnately compound, white dotted. Flowers sky-blue, blooming in Apr.-May; 3-18 flowers in a raceme. Fruits capsule, flat, pyramid-shape, narrowed on both ends, stigma at the tip. Seeds lustrous, arranged in two rows.

| Distribution | Endemic (Gwangwon-do, Gyeonggi-do, Gyeonsangbuk-do, Chungcheongbuk-do)

| Medicinal Name | Hyeon-ho-saek (玄胡索)

| Chemical components | Corydaline, dl-tetrahydropalmatine, protopine, l-tetrahydrocoptisine

| Effect | Menstrual pain, abdominal pain, lumbago, headache, neuralgia

| 이용부위 Parts Used |
덩이줄기 Tuber

현 호 색

Corydalis remota Fisch. ex Maxim.
현호색과 | Fumariaceae
Corydalis

| 식 물 | 전국의 산과 들에 자라는 여러해살이풀이다. 높이 20cm 정도로 식물체는 연약하다. 황색 덩이줄기에서 여린 줄기가 나와 곧게 서며 자란다. 1~2회 갈라지는 잎은 어긋나며 잎자루가 길다. 연한 홍자색 또는 청색인 꽃은 4~5월에 피고 5~10개가 원줄기 끝에 총상으로 달린다. 열매는 삭과이며 긴 타원형으로 양끝이 좁고 끝에 암술머리가 달려 있다. 종자는 흑색으로 매끄럽고 광택이 있다.

| 분 포 | 한국, 만주, 우수리

| 약 재 명 | 현호색(玄胡索)

| 성 분 | corydaline, dl-tetrahydropalmatine, protopine, l-tetrahydrocoptisine

| 효 능 | 생리통, 복통, 요통, 두통, 신경통

| Description | Perennial herb, widely distributed in mountains and open fields. Roots tuberous, yellow. Stems about 20cm tall, procumbent. Leaves alternate, 2-3 pinnately compound, long petiolate. Flowers light red-purple or blue, born in a raceme, blooming in Apr.-May. Fruits capsule, oblong, narrowed on both ends, stigma at the tip. Seeds black, glabrous, lustrous.

| Distribution | Korea, Manchuria, Ussuri

| Medicinal Name | Hyeon-ho-saek(玄胡索)

| Chemical components | Corydaline, dl-tetrahydropalmatine, protopine, l-tetrahydrocoptisine

| Effect | Menstrual pain, abdominal pain, lumbago, headache, neuralgia

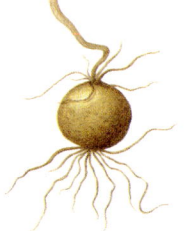

| 이용부위 **Parts Used** |
덩이줄기 Tuber

뱀딸기 *Duchesnea indica* (Andr.) Focke
장미과 | Rosaceae

| 식 물 | 전국의 밭이나 논둑의 양지에 흔하게 자라는 여러해살이풀이다. 덩굴이 옆으로 벋으면서 마디에서 뿌리가 내린다. 잎은 어긋나고 난형 또는 난상 원형의 소엽이 3장씩 달린다. 꽃은 4~5월에 노란색으로 피고 엽액에서 자라는 긴 꽃자루에 1개씩 달린다. 열매는 둥글고 연한 홍백색 바탕에 붉은빛이 도는 수과가 점처럼 흩어져 있다. |

| 분 포 | 한국, 일본, 중국, 만주, 말레이시아, 인도 |

| 약 재 명 | 사매(蛇莓) |

| 성 분 | linoleic acid, β-sitosterol, flavonoid, saponin |

| 효 능 | 구내염, 인후염, 종기, 디프테리아, 습진, 화상, 유방염, 타박상, 해수, 백일해, 코피, 토혈, 각혈, 자궁출혈, 이질 |

| Description | Perennial herb, widely distributed in open sunny areas of cultivated fields or adjacent banks. Stems trailing, rooting at nodes. Leaves alternate, three leaflets ovate or ovate-circular. Flowers yellow, solitary at the tip of long pedicel, blooming in Apr.-May. Fruits achene, red-tinged in red white, round, scattered like small dots. |

| Distribution | Korea, Japan, China, Manchuria, Malaysia, India |

| Medicinal Name | Sa-mae(蛇莓) |

| Chemical components | Linoleic acid, β-sitosterol, flavonoid, saponin |

| Effect | Stomatitis, sore throat, boil, diphtheria, eczema, burn, mastitis, bruise, cough, whooping cough, nosebleed, hematemesis, hemoptysis, uterine hemorrhage, dysentery |

| 이용부위 Parts Used |
지상부 Above-ground parts

큰괭이밥

Oxalis obtriangulata Maxim.
괭이밥과 | Oxalidaceae
Mountain Lady's-sorrel

식 물	산지의 숲 속에 나는 여러해살이풀이다. 뿌리줄기는 가늘게 옆으로 벋고 끝에 인편이 밀생한다. 잎은 긴 잎자루 끝에서 도삼각형의 작은 잎 3장이 옆으로 퍼져 달린다. 흰색 꽃은 5~6월에 피고 잎자루 사이에서 길게 발달하는 꽃줄기 끝에 1개씩 달린다. 열매는 삭과로 원주상 난형이고 익으면 5조각으로 벌어진다.
분 포	한국, 일본, 만주, 우수리
약재명	초장초(酢漿草)
성 분	초산염, malic acid, tartaric acid
효 능	설사, 이질, 치질, 황달, 소변불통, 인후염, 유방염, 종기, 옴, 버짐, 악창, 타박상

Description	Perennial herb, widely distributed in forests. Root stocks slender, rhizomatous; scales fasciculate on the end. Leaves 3-foliate obdeltoid leaflets, long-petioled. Flowers white, solitary at the tip of long pedicel, blooming in May-Jun. Fruits capsule, cylindrical ovate, crack opening into five valves after ripening.
Distribution	Korea, Japan, Manchuria, Ussuri
Medicinal Name	Cho-jang-cho(酢漿草)
Chemical components	Acetate, malic acid, tartaric acid
Effect	Diarrhoea, dysentery, hemorrhoids, jaundice, urination disorder, sore throat, mastitis, boil, scabies, trichophytia, obstinate abscess, bruise

| 이용부위 Parts Used |
지상부 Above-ground parts

다 래

Actinidia arguta (Siebold & Zucc.) Planch. ex Miq.
다 래 나 무 과 | Actinidiaceae
Vine Pear, Bower Actinidia, Tara Vine, Yang-tao

식 물	전국의 산지 숲 속에 자라는 낙엽성 덩굴나무이다. 뿌리는 사방으로 번는다. 줄기는 갈색으로 주위의 나무들을 감고 올라간다. 잎은 어긋나고 타원형 또는 넓은 타원형으로 가장자리에 잔톱니가 있다. 꽃은 5~6월에 녹백색으로 피고 암수딴그루로 엽액에 달린다. 열매는 장과로 난상 원형이고 10월에 황록색으로 익는다.
분 포	한국, 일본, 만주, 우수리, 사할린
약 재 명	미후리(獼猴梨)
성 분	당, vitamin, 유기산, 색소
효 능	소갈증, 급성전염성간염, 식욕부진, 소화불량

Description	Deciduous twining vine, dioecious, widely distributed in forests or forest edges. Roots stretching out. Stems brownish, curling up. Leaves alternate, elliptical or broadly elliptical, serrate. Flowers in an axil, dioecious, greenish white, blooming in May-Jun. Fruits berry, ovate round, yellowish green, ripening in Oct.
Distribution	Korea, Japan, Manchuria, Ussuri, Sakhalin
Medicinal Name	Mi-hu-ri(獼猴梨)
Chemical components	Sugar, vitamin, organic acid, pigment
Effect	Diabetes mellitus, acute infectious hepatitis, appetite impaired, dyspepsia

| 이용부위 Parts Used |
열매 Fruit

제비꽃

Viola mandshurica W.Becker
제비꽃과 | Violaceae

Manchurian Violet

| 식 물 | 전국의 들판에서 흔히 자라는 여러해살이풀이다. 뿌리는 몇 개로 갈라지며 짙은 갈색으로 짧은 뿌리줄기가 있다. 잎은 뿌리에서 뭉쳐나고 끝이 둔한 피침형이다. 보라색 또는 짙은 자주색 꽃은 4~5월에 피며 꽃줄기 끝에 1개씩 옆을 향해 달린다. 열매는 삭과로 익으면 3갈래로 벌어진다.
| 분 포 | 한국, 일본, 만주, 중국, 아무르, 우수리
| 약재명 | 자화지정(紫花地丁)
| 성 분 | saponin, flavonoid, cerotic acid, 불포화 지방산
| 효 능 | 종기, 발진, 맹장염, 결핵, 해열, 소염

| Description | Perennial herb, widely distributed in open fields. Roots numerous branched, dark brown, fibrous. Leaves radical, clustered, lanceolate, obtuse at the tip. Flowers violet or dark violet, solitary at the tip of long pedicel, blooming in Apr.-May. Fruits capsule, crack opening into three valves (parietal placentation).
| Distribution | Korea, Japan, Manchuria, China, Amur, Ussuri
| Medicinal Name | Ja-hwa-ji-jeong(紫花地丁)
| Chemical components | Saponin, flavonoid, cerotic acid, unsaturated fatty acid
| Effect | Boil, rash, appendicitis, tuberculosis, antipyretic, antiinflammation

| 이용부위 Parts Used |
지상부 Above-ground parts

가시오갈피

Eleutherococcus senticosus (Rupr. & Maxim.) Maxim.
두릅나무과 | Araliaceae
Siberian ginseng

식 물	지리산 이북의 깊은 산 숲 속에서 드물게 자라는 낙엽성 관목이다. 높이는 2~3m 정도이며 회갈색 줄기에는 가늘고 긴 가시가 빽빽하게 난다. 잎은 3~5개가 모인 장상엽으로 어긋나게 달리고 가장자리에 뾰족한 톱니가 있다. 자황색 꽃은 7월에 피며 가지 끝이나 아랫부분에 달린다. 열매는 둥글고 9월에 검은색으로 익는다.
분 포	한국, 일본, 사할린, 중국 동북부, 우수리
약재명	자오가(刺五加)
성 분	eleutheroside A, triterpenoid saponin Ⅰ·Ⅱ, protoprimulagenin A
효 능	식욕부진, 허약증, 요통, 무력증, 보행장애, 무릎관절염, 신경쇠약

Description	Deciduous shrub, rare in deep mountains north of Mt. Jiri. Commonly cultivated as medicinal plant. Stems 2-3m tall, grey brown, densely spiny. Leaves 3-5 palmate, alternate, serrate. Flowers purplish yellow, born in an umbel at the tip of stem or beneath, blooming in Jul. Fruits round, black, ripening in Sep.
Distribution	Korea, Japan, Sakhalin, Northeast China, Ussuri
Medicinal Name	Ja-o-ga(刺五加)
Chemical components	Eleutheroside A, triterpenoid saponin Ⅰ·Ⅱ, protoprimulagenin A
Effect	Appetite impaired, asthenic diathesis, lumbago, asthenia, walking difficulty, knee arthritis, neurasthenia

| 이용부위 Parts Used |
뿌리, 나무껍질
Root and rhytidome

인 삼

Panax ginseng C.A.Mey.
두릅나무과 | Araliaceae
Ginseng

| 식 물 | 깊은 산의 숲 속에 드물게 자라는 여러해살이풀이다. 흔히 농가에서 약용식물로 재배한다. 식물체는 60cm 정도까지 자라고 줄기는 매년 1개가 곧게 자란다. 연한 녹색의 꽃은 4월에 줄기 끝에 달린다. 열매는 납작한 구형으로 선홍색으로 익는다.

| 분 포 | 한국, 만주, 연해주

| 약재명 | 인삼(人蔘), 산삼(山蔘), 장뇌(樟腦) 잎-인삼엽(人蔘葉), 노두(蘆頭)-인삼로(人蔘蘆), 잔뿌리-미삼(尾蔘), 종자-인삼자(人蔘子), 찐것-홍삼(紅蔘)

| 성 분 | ginsengsaponin (gingenoside Ra · Rb1 · Rb2 · Rb3 · Rc · Rd · Re · R0), amino acid, 당류, vitamin, flavonoid, 휘발성 정유, 무기질

| 효 능 | **인삼**-신체허약, 권태, 피로, 다한증, 식욕부진, 구토, 설사, 소갈증, 건망증, 정신력 증강, 강장효과 **인삼엽**-폐열, 갈증 **인삼로**-설사, 원기회복 **미삼**-원기상승, 구토, 구역, 갈증 **인삼자**-종기 **홍삼**-혈액순환개선, 혈전형성억제, 면역기능활성화, 항노화작용

| 이용부위 **Parts Used** |
전초 All parts

| Description | Perennial herb, rare in deep mountains, mainly cultivated for medicinal purposes. Stem up to 60cm tall, solitary, erect. Flowers light green, born in an umbel at the tip of stem. Fruits globose, brightly red, ripening in Apr.

| Distribution | Korea, Manchuria, Maritime Provinces.

| Medicinal Name | In-sam(人蔘), san-sam(山蔘), jang-noe(樟腦); leaf - in-sam-yeop(人蔘葉); radix head - in-sam-ro(人蔘蘆); tertiary roots - mi-sam(尾蔘); seed - in-sam-ja (人蔘子); Steamed - hong-sam(紅蔘)

| Chemical components | Ginsengsaponin (gingenoside Ra·Rb1·Rb2·Rb3·Rc·Rd·Re·R0), amino acid, sugar, vitamin, flavonoid, volatile essential oil, minerals

| Effect | In-sam - asthenic diathesis, fatigue, tiredness, excessive sweating, appetite impaired, vomiting, diarrhoea, diabetes mellitus, forgetfulness, improvement mental health, tonic effect; In-sam-yeop - pyeyeol(fever skin), thirst; In-sam-ro - diarrhoea, energy recovery; Mi-sam - energy enhancement, vomiting, nausea, thirst; In-sam-ja - boil; Hong-sam - improvement blood circulation, platelet aggregation inhibition, immunity activation, antiaging

섬시호 *Bupleurum latissimum* Nakai
산형과 | Apiaceae

|식 물| 울릉도 해안의 숲 속이나 절벽에 드물게 자라는 여러해살이풀이다. 줄기는 높이 60cm에 달하고 털이 없으며 세로로 능선이 있다. 잎은 난형 또는 넓은 난형으로 가장자리가 물결 모양이고 잎자루는 위로 갈수록 짧아지거나 없어진다. 꽃은 5~6월에 피고 노란색으로 줄기나 가지 끝의 겹산형화서에 달린다. 열매는 분과로 난형이다.

|분 포| 특산(울릉도)

|약재명| 시호(柴胡)

|성 분| bupleurumol, oleic acid, saikosaponin A·B·C, saikogenin F·E·G

|효 능| 해열, 진정, 진통, 진해

|Description| Perennial herb, rare in forests or on cliffs along the seashore of Ulleung Island. Stems 60cm high, glabrous with longitudinal ridges. Leaves alternate, ovate or broadly ovate, wavy at the margin, petiole broadening downward. Flowers yellow, born in a compound umbel at the tip of stems or branches, blooming in May-Jun. Fruits ovate, schizocarp.

|Distribution| Endemic (Ulleungdo)

|Medicinal Name| Si-ho(柴胡)

|Chemical components| Bupleurumol, oleic acid, saikosaponin A·B·C, saikogenin F·E·G

|Effect| Antipyretic, sedation, pain relief, cough suppression

|이용부위 Parts Used|
뿌리 Root

산 수 유

Cornus officinalis Siebold & Zucc.
층층나무과 | Cornaceae
Japanese Cornelian Cherry, Japanese Cornel, Officinalis Dogwood

식 물	중부지방의 산에서 자라고 흔히 심어 기르는 낙엽성 아교목이다. 줄기는 갈색으로 비늘조각처럼 벗겨진다. 잎맥이 뚜렷한 잎은 마주나기로 달린다. 꽃은 이른 봄 3~4월경 잎보다 먼저 노란색으로 피며 산형화서에 달린다. 8~10월에 익는 선홍색 열매는 핵과로 광택이 있고 종자는 긴 타원형이며 능선이 있다.
분 포	한국, 중국
약 재 명	산수유(山茱萸)
성 분	cornin, verbenalin, tannin, saponin, ursolic acid, vitamin A
효 능	식은땀, 설사, 야뇨증, 자궁출혈, 이명

Description	Deciduous subtree, in mountains of the central peninsula, mainly cultivated. Stems brownish, scaly. Leaves opposite, with distinct veins. Flowers yellow, born in an umbel, blooming in the early spring, almost Mar.-Apr., before leaf emergence. Fruits drupe, red, lustrous, ripening in Aug.-Oct. Seeds oblong, with ridge.
Distribution	Korea, China
Medicinal Name	San-su-yu(山茱萸)
Chemical components	Cornin, verbenalin, tannin, saponin, ursolic acid, vitamin A
Effect	Cold sweat, diarrhoea, nocturia, uterine hemorrhage, sonitus

| 이용부위 Parts Used |
과육 Fruit

참좁쌀풀 *Lysimachia coreana* Nakai
앵초과 | Primulaceae
Korean Loosestrife

| 식 물 | 깊은 산의 양지바르고 습한 곳에 자라는 여러해살이풀이다. 뿌리줄기는 길게 옆으로 벋는다. 줄기는 곧추서며 가지가 갈라지기도 한다. 높이 50~100cm 정도이고 각이 지며 털은 거의 없다. 잎은 타원형 또는 난형이고 마주나거나 3장이 돌려나며 양면과 가장자리에 잔털이 있다. 꽃은 6~7월에 황색으로 피고 가지 끝이나 엽액에 달린다. 열매는 삭과로 둥글고 끝에 암술대가 남는다.

분 포 | 특산(함남, 강원, 경기, 경북, 충북)

약 재 명 | 황련화(黃蓮花)

효 능 | 고혈압

Description | Perennial herb, rare in open sunny and wet areas or deep mountain ridges. Root stocks rhizomatous. Stems erect, sometimes branched, 50-100cm tall, angled, glabrous. Leaves elliptical or ovate, opposite or 3-whorled, pubescent on both sides and margins. Flowers yellow, born at the tip of branches or at the leaf axils, blooming in Jun.-Jul. Fruits capsule, round, with a style on the tip.

Distribution | Endemic (Hamgyeongnam-do, Gwangwon-do, Gyeonggi-do, Gyeonsangbuk-do, Chungcheongbuk-do)

Medicinal Name | Hwang-nyeon-hwa(黃蓮花)

Effect | Hypertension

이용부위 Parts Used
지상부 Above-ground parts

큰까치수염

Lysimachia clethroides Duby
앵초과 | Primulaceae
Gooseneck Loosestrife

| 식 물 | 전국 산과 들의 양지바른 곳에 나는 여러해살이풀로 높이 약 1m 정도이다. 근경은 옆으로 벋고 줄기의 아래쪽은 약간 붉은 빛을 띤다. 잎은 어긋나며 가장자리가 밋밋하고 짧은 잎자루가 있다. 꽃은 백색으로 피고 6~7월에 원줄기 끝에서 한쪽으로 굽은 총상화서에 밀집하여 달린다. 화서는 길이 10~20cm이지만 결실기에는 길이가 40cm에 이르기도 한다. 꽃받침으로 싸여 있는 열매는 삭과로 구형이다.

| 분 포 | 한국, 일본, 중국, 러시아 극동부

| 약재명 | 진주채(珍珠菜)

| 성 분 | saponin, primulagenin A, dihydropriverogenin A

| 효 능 | 생리불순, 백대하, 이질, 인후염, 유방염

| Description | Perennial herb, widely distributed in open sunny places of mountains and open fields, 1m tall. Roots rhizomatous. Stems slightly red on the lower part. Leaves alternate, with a slender and short petiole. Flowers white, born in raceme in cluster at the tip of the main stem, blooming in Jun.-Jul. Inflorescence raceme, 10-20cm long, up to 40cm in the fruiting season. Fruits capsule, globose, enclosed within calyx.

| Distribution | Korea, Japan, China, Far Eastern Russia

| Medicinal Name | Jin-ju-chae(珍珠菜)

| Chemical components | Saponin, primulagenin A, dihydropriverogenin A

| Effect | Menstrual irregularity, leukorrhoea, dysentery, sore throat, mastitis

| 이용부위 Parts Used |
뿌리, 지상부
Root and above-ground parts

용담

Gentiana scabra Bunge
용 담 과 | Gentianaceae
Gentian

| 식 물 | 전국 산지의 풀밭에 자라는 여러해살이풀이다. 뿌리줄기는 짧고 굵은 수염뿌리가 있다. 줄기는 높이 20~60cm 정도로 곧추서고 4개의 가는 줄이 있으며 종종 적자색을 띤다. 잎은 마주나고 3맥이 있으며 가장자리에 까칠까칠한 돌기가 있다. 종 모양의 꽃은 8~10월에 자주색으로 피고 윗부분의 엽액과 끝에 달린다. 열매는 삭과로 10~11월에 익는다.

| 분 포 | 한국, 만주, 아무르, 우수리, 사할린, 동부시베리아

| 약 재 명 | 용담초(龍膽草)

| 성 분 | gintianine, gentiapicrin, gentianose

| 효 능 | 황달, 이질, 음부가려움증, 대하, 습진, 두통, 인후통, 간기능보호, 담즙분비촉진, 이뇨작용, 혈압강하작용, 진정작용

| Description | Perennial herb, relatively rare in grassland of mountains. Root stocks short, thick, fibrous. Stems 20-60cm tall, erect, 4-striated, frequently red-purple. Leaves opposite, with three veins, serrate on margins. Flowers purple, bell-shaped, born at the tip of the stem or in the axil, blooming in Aug.-Oct. Fruits capsule, ripening in Oct.-Nov.

| Distribution | Korea, Manchuria, Amur, Ussuri, Sakhalin, Eastern Siberia.

| Medicinal Name | Yong-dam-cho(龍膽草)

| Chemical components | Gintianine, gentiapicrin, gentianose

| Effect | Jaundice, dysentery, vaginal itching, leukorrhea, eczema, headache, sore throat, hepatoprotection, promotion bile secretion, diuretic effect, blood pressure lowering, sedation

| 이용부위 **Parts Used** |
뿌리 Root

2004. SEUNG-HYUN YI.

과 남 풀

Gentiana triflora var. *japonica* (Kusn.) H.Hara
용담과 | Gentianaceae
Threeflower Gentian

| 식 물 | 깊은 산의 습지에 자라는 여러해살이풀이다. 뿌리줄기는 짧고 굵은 수염뿌리가 있다. 줄기는 곧추서며 높이 30~80cm 정도이다. 잎은 마주나며 3~4맥이 있고 가장자리는 매끈하다. 종 모양의 꽃은 8~10월에 하늘색 또는 자주색으로 피고 윗부분의 엽액과 끝에 달린다. 열매는 삭과로 좁고 길며 2개로 갈라진다.
| 분 포 | 한국, 만주, 아무르, 우수리, 사할린, 동부시베리아
| 약재명 | 용담초(龍膽草)
| 성 분 | gintianine, gentiapicrin, gentianose
| 효 능 | 황달, 이질, 음부가려움증, 대하, 습진, 간기능보호, 담즙분비촉진, 이뇨작용, 혈압강하작용, 진정작용

| Description | Perennial herb, relatively rare in wet or sunny places of deep mountains. Root stocks short, thick, fibrous. Stems erect, 30-80cm tall. Leaves opposite, entire, with 3 or 4 veins. Flowers bell-shaped, sky-blue or purple, born at the tip of the stem or in the axil, blooming in Aug.-Oct. Fruits capsule, narrow, crack opening in two valves.
| Distribution | Korea, Manchuria, Amur, Ussuri, Sakhalin, Eastern Siberia
| Medicinal Name | Yong-dam-cho(龍膽草)
| Chemical components | Gintianine, gentiapicrin, gentianose
| Effect | Jaundice, dysentery, vaginal itching, leukorrhea, eczema, hepatoprotection, promotion bile secretion, diuretic effect, blood pressure lowering, sedation

| 이용부위 **Parts Used** |
뿌리 Root

큰구슬붕이

Gentiana zollingeri Faw.
용 담 과 | Gentianaceae
Zollinger Gentian

식 물	전국의 산과 들의 양지바른 곳에 나는 두해살이풀이다. 뿌리는 가늘고 길게 자란다. 줄기는 곧추서고 높이 5~10cm이며 능선과 잔돌기가 있다. 잎은 마주나며 난형 또는 넓은 난형이다. 꽃은 5~6월에 종 모양으로 피며 자줏빛이 돌고 원줄기 또는 가지 끝에 몇 개씩 모여 달린다. 열매는 삭과로 긴 자루가 있으며 익으면 2개로 갈라진다.
분 포	한국, 일본, 만주, 중국, 아무르, 우수리, 사할린
약 재 명	석용담(石龍膽)
효 능	충수염, 결핵성림프선염, 종기, 악창, 안구충혈

Description	Biennial, relatively rare in open sunny places of mountains and open fields. Roots long, slender. Stems erect, 5-10cm tall, with ridges, with small projections. Leaves opposite, ovate or broadly ovate. Flowers purple, bell-shaped, born at the tip of the main stem or branches in cluster, blooming in May-Jun. Fruits capsule, with a long stalk, crack opening in two valves after ripening.
Distribution	Korea, Japan, Manchuria, China, Amur, Ussuri, Sakhalin
Medicinal Name	Seog-yong-dam(石龍膽)
Effect	Appendicitis, tuberculous lymphadenitis, boil, obstinate abscess, bloodshot eyes

| 이용부위 Parts Used |
지상부 Above-ground parts

박 주 가 리

Metaplexis japonica (Thunb.) Makino
박 주 가 리 과 | Asclepiadaceae
Japanese Metaplexis

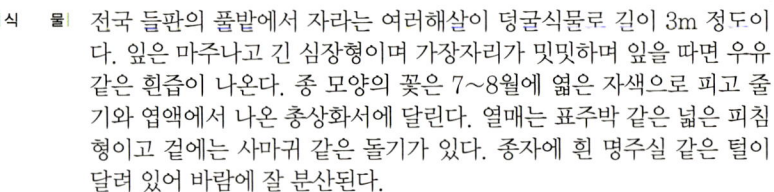

식 물	전국 들판의 풀밭에서 자라는 여러해살이 덩굴식물로 길이 3m 정도이다. 잎은 마주나고 긴 심장형이며 가장자리가 밋밋하며 잎을 따면 우유 같은 흰즙이 나온다. 종 모양의 꽃은 7~8월에 엷은 자색으로 피고 줄기와 엽액에서 나온 총상화서에 달린다. 열매는 표주박 같은 넓은 피침형이고 겉에는 사마귀 같은 돌기가 있다. 종자에 흰 명주실 같은 털이 달려 있어 바람에 잘 분산된다.
분 포	한국, 일본, 중국
약 재 명	라마(蘿藦)
성 분	뿌리-benzoylramanone, metaplexigenin, isoramanone 잎-d-cymorose, digitoxose
효 능	발기부전, 허약성대하, 유즙분비부족, 단독, 종기, 뱀에 물린 곳

Description	Perennial vine, widely distributed in grassland of open fields, 3m tall. Leaves opposite, long-cordate, entire, oozing white latex like milk when cut. Flowers pale purple, bell-shaped, born in a raceme from stems or leaf axils, blooming in Jul.-Aug. Fruits broadly lanceolate like a gourd dipper, with wart-like small protuberance on the surface. Seeds with silky hairs, dispersed by wind.
Distribution	Korea, Japan, China
Medicinal Name	Ra-ma(蘿藦)
Chemical components	Root - benzoylramanone, metaplexigenin, isoramanone Leaf - d-cymorose, digitoxose
Effect	Impotence, asthenic leukorrhea, lactation insufficiency, erysipelas, boil, snakebite

이용부위 Parts Used
뿌리, 지상부 Root and above-ground parts

누리장나무

Clerodendrum trichotomum Thunb.
마편초과 | Verbenaceae
Glory Bower, Harlequin Glory-bower

식 물	중부 이남의 산기슭이나 골짜기에 자라는 관목으로 높이 3m에 이른다. 줄기 전체에서 누린내가 나고 나무껍질은 잿빛이다. 잎은 마주나고 난형으로 끝이 뾰족하며 잎자루는 길이 3~10cm 정도이다. 꽃은 새 가지 끝에서 8~9월에 피고 흰색이며 꽃받침은 붉은색이다. 푸른 빛을 띤 보라색 열매는 핵과로 10월에 익으며 꽃받침에 싸인다.
분 포	한국, 일본, 대만, 중국
약재명	취오동(臭梧桐)
성 분	clerodendrin, mesoinositol, alkaloid, acacetin-7-glycurono-glucoronide
효 능	관절염, 사지마비, 반신불수, 습진, 피부가려움증, 고혈압, 편두통

Description	Deciduous shrub, widely distributed in the foothills of mountains or valleys south of the central peninsula, 3m tall. Stems stench. Bark ash-colored. Leaves opposite, ovate, acuminate, petiole 3-10cm long. Flowers white, born at the tip of new branches, blooming in Aug.-Sep., calyx is red. Fruits drupe, bluish purple, enclosed within calyx, ripening in Oct.
Distribution	Korea, Japan, Taiwan, China
Medicinal Name	Chwi-o-dong(臭梧桐)
Chemical components	Clerodendrin, mesoinositol, alkaloid, acacetin-7-glycurono-glucoronide
Effect	Arthritis, quadriplegia, hemiplegia, eczema, pruritus, hypertension, migraine

| 이용부위 **Parts Used** |
가지, 잎 Branch and Leaf

꽃향유

Elsholtzia splendens Nakai
꿀풀과 | Lamiaceae
Haichow Elsholtzia

식 물	거의 전국의 산과 들에 자라는 한해살이풀이다. 몇 개의 뿌리가 사방으로 벋고 여기에서 다시 잔뿌리가 나온다. 줄기는 곧추서나 가지를 많이 치고 네모가 지며 자색을 띤다. 잎은 마주나며 난형이고 가장자리에 톱니가 있다. 꽃은 9~10월에 피고 분홍빛이 나는 자주색이며 많은 꽃이 빽빽하게 한쪽으로 치우쳐서 달린다. 열매는 작은 견과로 좁은 도란형이다.
분 포	한국, 만주, 우수리
약재명	향유(香薷)
성 분	휘발성 정유
효 능	여름감기, 이뇨작용

Description	Annual, widely distributed in mountains and fields. Roots fibrous, creeping. Stems erect, many branched, 4-angled, purple. Leaves opposite, ovate, serrate. Flowers pink-colored purple, born in a cluster, blooming in Sep.-Oct. Fruits small nut, narrowly obovate.
Distribution	Korea, Manchuria, Ussuri
Medicinal Name	Hyang-yu(香薷)
Chemical components	Volatile essential oil
Effect	Summer cold, diuretic effect

| 이용부위 Parts Used |
지상부 Above-ground parts

벌깨덩굴

Meehania urticifolia (Miq.) Makino
꿀풀과 | Lamiaceae
Nettleleaf Meehania

식 물	전국의 산지 그늘진 곳에 자라는 여러해살이풀이다. 줄기는 네모지며 꽃이 진 뒤에 가지가 옆으로 벋으면서 마디에서 뿌리가 내린다. 잎은 마주나고 난상 심장형이며 가장자리에 둔한 톱니가 있다. 꽃은 5월에 자줏빛으로 피고 윗부분의 엽액에서 한쪽을 향해 2~6개가 달린다. 열매는 작은 견과로 도란형이다.
분 포	한국, 일본, 만주, 중국, 우수리
약재명	미한화(美漢花)
효 능	해열, 통증

Description	Perennial herb, widely distributed in shady places of mountains. Stems 4-angled, branched sideways, bearing roots from nodes after flower senescence. Leaves opposite, ovate-cordate, obtusely spinulous. Flowers purple, lateral, born on the upper part of axils with 2-6, blooming in May. Fruits small nut, obovate.
Distribution	Korea, Japan, Manchuria, China, Ussuri
Medicinal Name	Mi-han-hwa(美漢花)
Effect	Antipyretic, pain

| 이용부위 Parts Used |
지상부 Above-ground parts

백리향

Thymus quinquecostatus Celak.
꿀풀과 | Lamiaceae
Thyme, Fiveribbed Thyme

| 이용부위 **Parts Used** |
| 지상부 Above-ground parts |

식 물	높은 산의 산정이나 바닷가 바위틈에서 드물게 자라는 낙엽성 반관목으로 높이는 3~15cm 정도이다. 땅 위에 퍼져 자라고 어린 가지는 비스듬히 서며 향기가 난다. 잎은 마주나고 긴 타원형으로 양면에 선점이 있으며 가장자리는 밋밋하거나 물결 모양의 톱니가 있다. 꽃은 6~8월에 엽액에서 2~4개씩 달리지만 가지 끝에 모여 수상화서처럼 보인다. 분홍색 또는 드물게 흰색이다. 열매는 작은 견과로 9월에 짙은 갈색으로 익는다.
분 포	한국, 일본, 만주, 중국, 몽골, 인도
약재명	백리향(百里香)
성 분	scutellareinheteroside, Luteolin-7-glucoside, apigenin, carvacrol
효 능	구토, 복통, 설사, 복부창만, 기침, 백일해, 인후염, 관절염, 동통

Description	Deciduous subshrub, uncommon in the summit of high mountains or rock crevices along seashore, 3-15cm tall, creeping. Young branches erect on the slant, fragrant. Leaves opposite, elliptical, with pellucid dots on both sides, entire or spinulose in the wavy pattern. Flowers pink or infrequently white, born in axils with 2-4, in cluster on the tip of branches like spite, blooming in Jun.-Aug. Fruits small nut, dark brown, ripening in Sep.
Distribution	Korea, Japan, Manchuria, China, Mongolia, India
Medicinal Name	Baeng-ni-hyang(百里香)
Chemical components	Scutellareinheteroside, Luteolin-7-glucoside, apigenin, carvacrol
Effect	Vomiting, abdominal pain, diarrhoea, abdominal fullness, cough, whooping cough, sore throat, arthritis, pain

섬백리향

Thymus quinquecostatus var. *japonica* Hara
꿀풀과 | Lamiaceae
Dagalet Thyme

| 식 물 | 울릉도 바닷가의 바위틈에서 자라는 낙엽성 반관목으로 높이는 3~15cm 정도이다. 원줄기는 땅 위로 퍼져 자라고 어린 가지가 비스듬히 서며 향기가 난다. 잎은 마주나고 긴 타원형으로 양면에 선점이 있으며 가장자리는 밋밋하거나 물결 모양의 톱니가 있다. 꽃은 6~8월에 엽액에서 2~4개씩 달리지만 가지 끝에 모여 수상화서처럼 보인다. 분홍색 또는 드물게 흰색이다. 열매는 작은 견과로 9월에 짙은 갈색으로 익는다.
_꽃과 잎의 길이가 백리향보다는 약간 길고 줄기가 더 굵다. |
분 포	울릉도
약재명	백리향(百里香)
성 분	scutellareinheteroside, Luteolin-7-glucoside, apigenin, carvacrol
효 능	구토, 복통, 설사, 복부창만, 기침, 백일해, 인후염, 관절염, 동통

이용부위 **Parts Used**
지상부 Above-ground parts

Description	Deciduous subshrub, rare in rock crevices along the Ulleung Island seashore, 3-15cm tall. Main stems creeping; young branches erect on the slant, fragrant. Leaves opposite, elliptical, with pellucid dots on both sides, entire or spinulose in the wavy pattern. Flowers pink or infrequently white, 2-4 flowers born in the leaf axils or in cluster on the tip of branches, blooming in Jun.-Aug. Fruits small nut, dark brown, ripening in Sep. Flowers and leaves longer, stems thicker than those of Baengnihyang.
Distribution	Ulleungdo
Medicinal Name	Baeng-ni-hyang(百里香)
Chemical components	Scutellareinheteroside, Luteolin-7-glucoside, apigenin, carvacrol
Effect	Vomiting, abdominal pain, diarrhoea, abdominal fullness, cough, whooping cough, sore throat, arthritis, pain

선갈퀴

Asperula odorata L.
꼭두서니과 | Rubiaceae
Woodruff

| 식 물 | 중부 이북과 울릉도의 산지 그늘진 곳에 자라는 여러해살이풀이다. 뿌리줄기는 옆으로 벋는다. 사각형의 줄기는 높이 25~40cm 정도이다. 엽병이 없는 긴 타원형의 잎은 6~10개가 돌려나고 잎 뒷면의 주맥과 가장자리에 위를 향한 거센털이 있다. 흰색의 꽃은 5~6월경에 줄기 끝에 취산화서로 달리고 화관은 깔때기 모양으로 끝이 4개로 갈라지며 수평으로 퍼진다. 열매는 둥글고 갈고리 모양의 털이 밀생한다.

| 분 포 | 한국, 일본, 사할린, 유럽, 북아메리카

| 약재명 | 육엽률(六葉葎)

| 효 능 | 전신부종, 피부궤양, 발진, 종기, 화상

| Description | Perennial herb, relatively rare in the shady places of mountains north of the central peninsula or in Ulleungdo. Root stocks rhizomatous. Stems rectangular, 25-45cm high. Leaves no petiole, elliptical, whorled with 6-10, with tough hairs upwards on the vein of the beneath and margins. Flowers white, born in cyme on the tip of stems, blooming in May-Jun. corolla funnel shaped, 4-segmented, spreading horizontally. Fruits round, with hooked hairs fasciculate.

| Distribution | Korea, Japan, Sakhalin, Europe, North Africa

| Medicinal Name | Yung-nyeop-ryul(六葉葎)

| Effect | Oedema generalized, skin ulcer, rash, boil, burn

| 이용부위 Parts Used |
지상부 Above-ground parts

인 동

Lonicera japonica Thunb.
인동과 | Caprifoliaceae
Japanese Honeysuckle, Gold-and-silver Flower

| 식 물 | 전국 산과 들의 양지바른 곳에 자라는 반상록성 덩굴식물이다. 줄기는 오른쪽으로 감겨 올라가고 붉은 빛이 도는 갈색이며 속은 비어 있다. 잎은 마주나며 긴 타원형이거나 넓은 피침형이다. 꽃은 6~7월에 피고 1~2개씩 엽액에 달리며 백색으로 피나 때때로 연한 붉은색을 띠고 나중에 노란색으로 변한다. 열매는 장과로 둥글며 10~11월에 검게 익는다.

| 분 포 | 한국, 일본, 만주, 중국

| 약재명 | 금은화(金銀花)-꽃, 인동등(忍冬藤)-줄기

| 성 분 | 금은화-luteolin, inositol, saponin, tannin
인동등-lonicerin, luteolin-7-rhamno-glucoside

| 효 능 | 금은화-여름 감기, 맹장염, 복막염, 자궁내막염, 이질, 유선염
인동등-해열, 전신통, 발한, 염증, 피부가려움증, 종기, 전염성간염, 근골동통, 풍습성관절염

| Description | Evergreen or semi-evergreen vine, widely distributed in open sunny places of mountains and fields. Stem reddish brown, hollow, winding up counter-clock-wise. Leaves opposite, elliptical or broadly lancelolate. Flowers white or frequently light red, turning yellow as mature, 1-2 flowers born in an axil, blooming in Jun-Jul. Fruits berry, round, black, ripening in Oct.-Nov.

| Distribution | Korea, Japan, Manchuria, China

| Medicinal Name | Geum-eun-hwa(金銀花) - flower; In-dong-deung(忍冬藤) - stem

| Chemical components | Geum-eun-hwa - luteolin, inositol, saponin, tannin;
In-dong-deung - lonicerin, luteolin-7-rhamno-glucoside

| Effect | Geum-eun-hwa - summer cold, appendicitis, peritonitis, endometritis, dysentery, mastitis; In-dong-deung - antipyretic, whole body pain, perspiration, inflammation, pruritus, boil, infectious hepatitis, musculoskeletal pain, rheumatic arthritis

| 이용부위 Parts Used |
꽃, 줄기 Flower and stem

도라지

Platycodon grandiflorum (Jacq.) A.DC.
초 롱 꽃 과 | Campanulaceae
Balloon Flower, Chinese Bellflower

| 식 물 | 전국의 산과 들에 자라는 여러해살이풀로 높이는 40~100cm 정도이다. 다육질의 뿌리는 굵게 자란다. 줄기는 회록색이며 직립하고 자르면 백색의 유액이 나온다. 잎은 마주나거나 돌려나고 또는 어긋나며 긴 난형 또는 넓은 피침형으로 가장자리에 톱니가 있다. 꽃은 7~8월에 피고 하늘색 또는 백색이며 원줄기 끝에 1개 또는 여러 개가 위를 향해 달린다. 열매는 삭과로 도란형이다.

| 분 포 | 한국, 일본, 중국, 아무르, 우수리, 다후리아

| 약 재 명 | 길경(桔梗)

| 성 분 | polygalaic acid, platycodigenin, platycogenic acid, platycodonin, platycodin A·B·C

| 효 능 | 해수, 가래, 코막힘, 두통, 편도선염, 인후염, 이질, 후중

| Description | Perennial herb, widely distributed in forests and grassy areas, commonly cuntivated. 40-100cm tall. Roots fleshy, thickening as mature. Stems grayish green, erect, oozing white latex like milk when cut. Leaves opposite or whorled, long-ovate or ovate-lanceolate, spinulose on margins. Flowers sky-blue or white, upward, 1-several flowers born at the tip of the main stem, blooming in Jul.-Aug. Fruits capsule, obovate.

| Distribution | Korea, Japan, China, Amur, Ussuri, Dahuria

| Medicinal Name | Gil-gyeong(桔梗)

| Chemical components | Polygalaic acid, platycodigenin, platycogenic acid, platycodonin, platycodin A·B·C

| Effect | Cough, sputum, nasal stuffiness, headache, tonsillitis, sore throat, dysentery, incomplete defecation

| 이용부위 Parts Used |
뿌리 Root

벌개미취

Aster koraiensis Nakai
국화과 | Asteraceae
Korean Starwort, Korean Daisy

식 물	산지의 양지바른 곳 또는 습지에 자라는 여러해살이풀이다. 근경은 적갈색으로 옆으로 벋으며 잔뿌리가 많고 식물체는 높이 50~60cm 정도이다. 줄기는 곧게 자라고 세로로 파인 홈과 줄이 있다. 잎은 어긋나며 피침형이나 위로 갈수록 점차 작아져서 선형으로 된다. 꽃은 6~10월에 피고 연한 자주색이며 가지 끝과 원줄기 끝에 달린다. 열매는 수과로 11월에 익는다.
분 포	특산(경기도 이남)
이용부위	뿌리
약 재 명	자원(紫菀)
성 분	saponin, epifriedelin, friedelin, shionone, aster-saponin
효 능	해수, 가래, 천식, 항균작용

Description	Perennial herb, relatively rare in open sunny fields or wet places of mountains, commonly cuntivated as landscaping and holticultural plant. 50-60cm tall. Roots red brown, rhizomatous, fibrous. Stems erect, vertically glandular, grooved. Leaves alternate and lanceolate, attenuate upwards, finally linear. Flowers light purple, the heads (capitula) born at the tip of branches or main stems, blooming in Jun.-Oct. Fruits achene, ripening in Nov.
Distribution	Endemic (South of Gyeonggi-do)
Medicinal Name	Ja-won(紫菀)
Chemical components	Saponin, epifriedelin, friedelin, shionone, aster-saponin
Effect	Cough, sputum, asthma, antibiotic effect

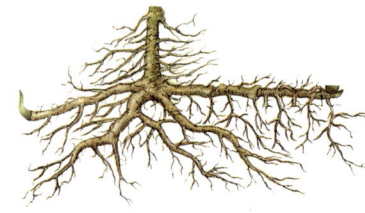

| 이용부위 Parts Used |
뿌리 Root

삽주

Atractylodes ovata (Thunb.) DC.
국화과 | Asteraceae
Japanese Atracty-lodes

| 식 물 | 전국 산지의 건조한 곳에 자라는 여러해살이풀이다. 굵고 긴 뿌리는 마디가 있고 향기가 있다. 줄기의 윗부분에서 가지가 몇 개 갈라지며 높이는 30~100cm 정도이다. 잎은 어긋나고 줄기 밑 부분에 달린 잎은 새깃 모양으로 깊게 3~5개로 갈라진다. 잎 표면은 광택이 있고 뒷면에 흰빛이 돌며 가장자리에 가시 같은 톱니가 있다. 줄기 윗부분에 달린 잎은 갈라지지 않고 잎자루가 거의 없다. 꽃은 7~10월에 흰색으로 피며 줄기와 가지 끝에 두상화가 1개씩 달린다. 포는 꽃과 길이가 같고 2줄로 달리며 깃꼴로 갈라진다. 열매는 수과로 털이 있으며 갈색 관모가 있다. |

분 포	한국, 일본, 중국 동북부
약재명	창출(蒼朮), 백출(白朮)
성 분	정유, atractylon, actractylol, vitamin A · B
효 능	식욕부진, 소화불량, 위장염, 감기, 야맹증, 이뇨작용, 소염작용

이용부위 **Parts Used**
뿌리 Root

Description	Perennial herb, widely distributed in dry places of mountains. Roots thick, long, fragrant. Stem branched on the upper, 30-100cm high. Leaves alternate. Lower cauline leaves pinnately 3 to 5-lobed; the surface lustrous, beneath tinged with white. Upper cauline leaves not lobed and subsessile. Flowers in head, white, a sinlge head (capitulum) born at the tip of branches or stems, blooming in Jul.-Oct. Bracteoles same as flower in length, 2-seriate, bipinnately segmented. Fruits achene, hairy, with brownish pappus.
Distribution	Korea, Japan, Northeast China
Medicinal Name	Chang-chul(蒼朮),, baek-chul(白朮)
Chemical components	Essential oil, atractylon, actractylol, vitamin A·B
Effect	Appetite impaired, dyspepsia, gastritis, cold, night blindness, diuretic effect, anti-inflammation

큰엉겅퀴

Cirsium pendulum Fisch. ex DC.
국 화 과 | Asteraceae
Pendulate Thistle

| 식 물 | 숲 가장자리와 강가의 습지 양지바른 곳에 자라는 여러해살이풀이다. 줄기는 곧게 서고 높이 1~2m 정도이다. 윗부분에서 가지가 많이 갈라지고 세로로 줄이 있으며 거미줄 같은 털이 있다. 근생엽과 줄기 기부에 달린 잎은 꽃이 필 때 말라 없어지며 잎 양면에 털이 있다. 줄기 중간에 달린 잎은 어긋나고 끝이 길고 뾰족하며 잎자루가 없고 깃 모양으로 깊게 갈라진다. 자주색의 두상화는 7~10월에 피며 가지와 줄기 끝에서 밑을 향하여 달린다. 열매는 수과로 긴 타원형이며 4개의 모가 난 줄이 있고 백색 관모가 있다.

| 분 포 | 한국, 일본, 중국 동북부, 사할린, 시베리아 동부

| 약 재 명 | 대계(大薊)

| 성 분 | alkaloid, 정유

| 효 능 | 각혈, 코피, 자궁출혈, 소변출혈, 종기, 황달, 신경통

이용부위 **Parts Used**
뿌리, 지상부
Root and above-ground parts

| Description | Perennial herb, relatively common in open sunny places of forest edges or the wet land of river banks. Stem erect, 1-2m tall, many segmented on the upper part, vertically glandular, with cobweb-like hairs. Radical and basal leaves withering when flowers bloom, pubescent above and beneath. Stem leaves alternate, oblong, tipped, sessile, pinnately lobed. Flowers in head, purple, downwards, born at the tip of branches and stems, blooming in Jul.-Oct. Fruits achene, oblong, with 4-angled lines, with white pappus.

| Distribution | Korea, Japan, Northeast China, Sakhalin, Eastern Siberia

| Medicinal Name | Dae-gye(大薊)

| Chemical components | Alkaloid, essential oil

| Effect | Hemoptysis, nosebleed, uterine hemorrhage, hematuria, boil, jaundice, neuralgia

바늘엉겅퀴

Cirsium rhinoceros (H. Lév. & Vaniot) Nakai

국화과 | Asteraceae

| 식 물 | 제주도의 산지에 드물게 자라는 여러해살이풀이다. 뿌리는 양끝이 뾰족한 원기둥 모양이다. 줄기는 곧추서며 가지가 많이 갈라지고 세로줄과 털이 있다. 잎은 어긋나며 규칙적인 깃꼴로 갈라지고 가장자리에 딱딱하며 날카로운 가시가 있다. 자주색의 꽃은 7~8월에 줄기와 가지 끝에 달린다. 열매는 수과로 윗부분은 노란색이고 다른 부분은 자주색이며 관모는 갈색이다. |

분 포	특산(제주도)
약재명	대계(大薊)
성 분	alkaloid, 정유
효 능	각혈, 코피, 자궁출혈, 소변출혈, 종기, 황달, 고혈압, 신경통

Description	Perennial herb, relatively rare in the mountains of Jeju Island. Roots cylindrical, acuminate at tips. Stems erect, many branched, vertically glandular, hairy. Leaves alternate, pinnately lobed on a regular basis, with hard and pointed spines on margins. Flowers purple, heads born at the tip of stems or branches, blooming in Jul.-Aug. Fruits achene, yellow on the upper, purple on the rest; pappus brownish.
Distribution	Endemic (Jeju-do)
Medicinal Name	Dae-gye(大薊)
Chemical components	Alkaloid, essential oil
Effect	Hemoptysis, nosebleed, uterine hemorrhage, hematuria, boil, jaundice, hypertension, neuralgia

| 이용부위 **Parts Used** |
뿌리, 지상부
Root and above-ground parts

산국

Dendranthema boreale (Makino) Ling ex Kitam.
국화과 | Asteraceae

North Chrysanthemum

식 물	전국의 산지나 들의 양지바른 곳에 자라는 여러해살이풀이다. 뿌리줄기는 가늘고 길게 자란다. 줄기는 곧추서고 가지가 많이 갈라진다. 잎은 마주나며 깃꼴로 갈라지고 가장자리에 톱니가 있다. 꽃은 9~10월에 노란색으로 피고 가지와 줄기 끝에 다소 산형으로 달린다. 열매는 수과로 도란형이다.
분 포	한국, 일본, 만주, 중국
약재명	야국화(野菊花)
성 분	휘발성 정유, 다당류
효 능	종기, 인후염, 고혈압

Description	Perennial herb, widely distributed in open sunny places of mountains or fields. Root stocks slender, lengthy. Stems erect, many branched. Leaves opposite, pinnately lobed, spinulose on margins. Flowers yellow, heads born nearly in a corymb at the tip of stems or branches, blooming in Set.-Oct. Fruits achene, obovate.
Distribution	Korea, Japan, Manchuria, China
Medicinal Name	Ya-guk-hwa(野菊花)
Chemical components	Volatile essential oil, polysaccharides
Effect	Boil, sore throat, hypertension

| 이용부위 Parts Used |
꽃봉오리 Flower bud

좀씀바귀 *Ixeris stolonifera* A.Gray
국화과 | Asteraceae

| 식 물 | 전국 산지의 양지바른 곳에 자라는 여러해살이풀이다. 줄기는 포복성으로 땅 위를 기며 뿌리를 낸다. 잎은 어긋나고 넓은 난형 또는 넓은 타원형이다. 꽃은 5~6월에 노란색으로 피고 줄기 끝에 1~3개가 달린다. 열매는 좁은 방추형으로 관모는 흰색이다. |

| 분 포 | 한국, 일본, 중국 |

| 약 재 명 | 고채(苦菜) |

| 성 분 | inulin, synaroside, aliphatic, triterpenoid, sesquiterpene glicoside |

| 효 능 | 이질, 유방염, 구내염 |

| Description | Perennial herb, widely distributed in open sunny places of mountains. Stems procumbent, bearing roots. Leaves alternate, broadly ovate or broadly elliptical. Flowers yellow, 1-3 heads born at the tip of stems, blooming in May-Jun. Fruits narrowly pyramid-shaped; pappus white. |

| Distribution | Korea, Japan, China |

| Medicinal Name | Go-chae(苦菜) |

| Chemical components | Inulin, synaroside, aliphatic, triterpenoid, sesquiterpene glicoside |

| Effect | Dysentery, mastitis, stomatitis |

|이용부위 Parts Used|
지상부 Above-ground parts

왕씀배

Prenanthes ochroleuca (Maxim.) Hemsl.
국화과 | Asteraceae

| 식 물 | 제주도와 경기도의 산지 숲 속 습한 곳에 드물게 자라는 여러해살이풀이다. 뿌리는 굵다. 줄기는 곧추서며 높이 70~100cm 정도이다. 뿌리에 달린 잎은 꽃이 필 때 없어지고 줄기에 달린 잎은 어긋난다. 밑부분에 난 잎은 긴 잎자루에 달리며 크게 3개로 갈라지고 가장자리에 톱니가 있다. 위로 갈수록 잎자루가 없어지며 갈라지지 않고 가장자리가 밋밋하다. 꽃은 8~9월에 피며 노란색이다. 열매는 수과로 선형이고 관모는 검은빛이 도는 갈색이다. |

| 분 포 | 한국, 만주, 우수리 |

| 약재명 | 고채(苦菜) |

| 성 분 | inulin, synaroside, aliphatic, triterpenoid, sesquiterpene glicoside |

| 효 능 | 이질, 유방염, 구내염 |

| Description | Perennial herb, relatively rare in the wet places of the forest in Jeju Island and Gyeonggi-do. Roots thick. Stems erect, 70~100cm tall. Radical leaves withering when flowers bloom; cauline leaves alternate. Basal leaves long-petioled, 3-lobed, spinulose on margins; Upper leaves becoming sessile, entire. Flowers yellow, blooming in Aug.-Sep. Fruits achene, linear; pappus black-tinged brown. |

| Distribution | Korea, Manchuria, Ussuri |

| Medicinal Name | Go-chae(苦菜) |

| Chemical components | Inulin, synaroside, aliphatic, triterpenoid, sesquiterpene glicoside |

| Effect | Dysentery, mastitis, stomatitis |

| 이용부위 Parts Used |
지상부 Above-ground parts

미역취

Solidago virgaurea subsp. *asiatica* Kitam. ex Hara
국 화 과 | Asteraceae
Goldenrod

| 식 물 | 전국 산과 들의 양지바른 곳에 나는 여러해살이풀이다. 뿌리는 가는 수염뿌리가 사방으로 퍼진다. 줄기는 흔히 암자색을 띠고 곧게 서며 잔털이 있다. 잎은 난형, 긴 타원형, 또는 피침형으로 끝이 뾰족하며 가장자리에 톱니가 있고 위로 올라갈수록 좁아진다. 꽃은 7~10월에 노란색으로 피고 3~5개의 두상화가 산방화서를 이룬다. 열매는 수과로 원통형이다.
| 분 포 | 한국, 일본
| 약 재 명 | 일지황화(一支黃花)
| 성 분 | 정유, tannin, saponin, flavonoid, chlorogenic acid, caffeic acid
| 효 능 | 두통, 인후염, 편도선염, 황달, 타박상, 종기

| Description | Perennial herb, widely distributed in open sunny places of mountains and fields. Roots slender, fibrous, stretching. Stems dark purple, erect, with fine hairs. Leaves ovate, oblong or laneolate, acuminate at tips, spinulose on margins, narrowed as upwards. Flowers yellow, 3-5 heads born in a corymb, blooming in Jul.-Oct. Fruits achene, cylindrical.
| Distribution | Korea, Japan
| Medicinal Name | Il-ji-hwang-hwa(一支黃花)
| Chemical components | Essential oil, tannin, saponin, flavonoid, chlorogenic acid, caffeic acid
| Effect | Headache, sore throat, tonsillitis, jaundice, bruise, boil

| 이용부위 Parts Used |
지상부 Above-ground parts

흑삼릉 *Sparganium erectum* L.
흑삼릉과 | Sparganiaceae
Bur Reed, Knope-sedge

| 식 물 | 연못이나 도랑가의 얕은 물속에 드물게 자라는 여러해살이풀이다. 키는 1m 정도이며 옆으로 벋는 짧은 뿌리줄기가 있다. 줄기는 거칠고 강하다. 잎은 선형으로 뒷면 가운데에 맥이 튀어나와 있고 2열로 마주나며 아래 부분이 줄기를 감싼다. 꽃은 6~7월에 피고 흰색이며 엽액에서 둥그렇게 무리지어 두상화처럼 달린다. 이러한 두상화 여러 개가 차례로 매달려 하나의 꽃처럼 보이며 화서의 아래쪽에는 암꽃, 위쪽에는 수꽃이 핀다. 열매는 여러 개가 모여 둥그란 하나의 공처럼 익는다.
| 분 포 | 한국, 일본, 중국, 아프카니스탄
| 약재명 | 삼릉(三棱)
| 성 분 | 정유, 전분
| 효 능 | 월경폐색불통, 징가, 적취, 소화불량

| Description | Perennial herb, rare in swallow ponds or ditches, 1m tall. Root stocks short, stretching. Stems harsh, strong. Leaves linear, opposite in two rows, with protruded veins underneath, enclosing stems on the lower part. Flowers born in a spherical head, white, born in axils in round cluster blooming in Jun.-Jul. Female heads on the lower part of the peduncle, male heads above. Fruits ripening in cluster like a ball.
| Distribution | Korea, Japan, China, Afganistan
| Medicinal Name | Sam-neung(三棱)
| Chemical components | Essential oil, starch
| Effect | Dysmenorrhea, Jingga (genital tumors), Jeokchwi (tumors), dyspepsia

| 이용부위 Parts Used |
뿌리줄기 Root stock

둥근잎천남성 *Arisaema amurense* Maxim.
천남성과 | Araceae

|시 물| 전국의 산지에 나는 여러해살이풀이다. 덩이줄기는 편평한 구형으로 지름 2~4cm이고 주위에 작은 덩이줄기가 달려 번식한다. 윗부분에서 수염뿌리가 사방으로 퍼지고 얇은 막질의 인편이 겉을 감싸고 있다. 줄기는 곧추서고 녹색이다. 잎은 1~2장으로 3~5개의 소엽으로 갈라지며 가장자리는 매끈하다. 꽃은 5~7월에 피고 육수화서로 달린다. 불염포는 녹색이나 안쪽이 자색을 띠는 것도 있으며 윗부분이 모자처럼 앞으로 꼬부라지고 끝이 뾰족하다. 육수화서의 연장부는 곤봉 모양이다. 열매는 장과로 옥수수 모양이며 붉은색으로 익는다.

|분 포| 한국, 만주, 아무르, 우수리

|약 재 명| 천남성(天南星)

|성 분| triterphenesaponin, benzoic acid, 전분, amino acid

|효 능| 중풍, 반신불수, 구안와사, 수족마비, 전간, 파상풍, 자궁경부암

|Description| Perennial herb, widely distributed in mountains. Tubers depressed, globose, 2-4cm in diameter, with small ones on the upper part, fibrous roots stretching, thin membranous scales enclosing the surface. Stems erect, green colored. Leaves simple or double, lobed into 3-5 leaflets, entire on margins. Flowers born in a spadix, blooming in May-Jul. Spathe green, sometimes purple colored inside; the upper part bending forwards like a cap, acuminate, the extended part club-shaped. Fruits berry, corn-shaped, ripening in red.

|Distribution| Korea, Manchuria, Amur, Ussuri

|Medicinal Name| Cheon-nam-seong(天南星)

|Chemical components| Triterphenesaponin, benzoic acid, starch, amino acid

|Effect| Stroke, hemiplegia, Bell's palsy, numbness in the hands and feet, epilepsy, tetanus, cervical cancer

|이용부위 Parts Used| 덩이줄기 Tuber

천남성

Arisaema amurense for. *serratum* (Nakai) Kitagausa
천남성과 | Araceae
Serrate Amur Jackinthepulpit

| 식 물 | 전국의 산지에 나는 여러해살이풀이다. 덩이줄기는 편평한 구형으로 지름 2~4cm이고 주위에 작은 덩이줄기가 달려 번식한다. 윗부분에서 수염뿌리가 사방으로 퍼지고 얇은 막질의 인편이 겉을 감싸고 있다. 줄기는 곧추서고 녹색이다. 잎은 1~2장으로 3~5개의 소엽으로 갈라지며 가장자리에 톱니가 있다. 꽃은 5~7월에 피고 육수화서로 달린다. 불염포는 녹색이며 윗부분이 모자처럼 앞으로 꼬부라지고 끝이 뾰족하다. 육수화서의 연장부는 곤봉 모양이다. 열매는 장과로 옥수수 모양이며 붉은색으로 익는다. |

분 포	한국, 만주, 아무르, 우수리
약 재 명	천남성(天南星)
성 분	triterphenesaponin, benzoic acid, 전분, amino acid
효 능	중풍, 반신불수, 구안와사, 수족마비, 전간, 파상풍, 자궁경부암

Description	Perennial herb, widely distributed in mountains. Tubers depressed, globose, 2-4cm in diameter, with small ones on the upper part, fibrous roots stretching, thin membranous scales enclosing the surface. Stems erect, green colored. Leaves simple or double, lobed into 3-5 leaflets, serrate on margins. Flowers born in spadix, blooming in May-Jul. Spathes green the upper part bending forwards like a cap, acuminate, the extended part club-shaped. Fruits berry, ripening in red.
Distribution	Korea, Manchuria, Amur, Ussuri
Medicinal Name	Cheon-nam-seong(天南星)
Chemical components	Triterphenesaponin, benzoic acid, starch, amino acid
Effect	Stroke, hemiplegia, Bell's palsy, numbness in the hands and feet, epilepsy, tetanus, cervical cancer

이용부위 **Parts Used**
덩이줄기 Tuber

두루미천남성

Arisaema heterophyllum Blume
천남성과 | Araceae
Diversileaf Jackinthepulpit

| 식 물 | 전국의 산지에 나는 여러해살이풀이다. 편평한 구형의 덩이줄기는 지름 2~4cm이고 주위에 작은 덩이줄기가 달려 번식한다. 윗부분에서 수염뿌리가 사방으로 퍼지고 얇은 막질의 인편이 겉을 감싸고 있다. 줄기는 곧추서고 녹색이다. 잎은 1장으로 새가 날개를 펼친 것처럼 13~19개의 소엽으로 갈라진다. 꽃은 5~6월에 피고 육수화서로 달린다. 불염포는 녹색이며 윗부분이 모자처럼 앞으로 꼬부라지고 끝이 뾰족하다. 육수화서의 연장부는 채찍처럼 길게 자라 위로 선다. 열매는 장과로 옥수수 모양이며 붉은색으로 익는다. |

분 포	한국, 일본, 대만, 중국
약 재 명	천남성(天南星)
성 분	triterphenesaponin, benzoic acid, 전분, amino acid
효 능	중풍, 반신불수, 구안와사, 수족마비, 전간, 파상풍, 자궁경부암

이용부위 **Parts Used**
덩이줄기 Tuber

Description	Perennial herb, rare in mountains. Tubers depressed, globose, 2-4cm in diameter, with small ones on the upper part, fibrous roots stretching, thin membranous scales enclosing the surface. Stems erect, green colored. Leaves compound with 13-19 leaflets as a bird spreading wings. Flowers born in a spadix, blooming in May-Jul. Spathe green; the upper part bending forwards like a cap, acuminate, the extended part stretching upward like a whip. Fruits berry, corn-shaped, ripening in red.
Distribution	Korea, Japan, Taiwan, China
Medicinal Name	Cheon-nam-seong(天南星)
Chemical components	Triterphenesaponin, benzoic acid, starch, amino acid
Effect	Stroke, hemiplegia, Bell's palsy, numbness in the hands and feet, epilepsy, tetanus, cervical cancer

점박이천남성 *Arisaema peninsulae* Nakai
천 남 성 과 | Araceae
Korean Jackinthepulpit

| 식 물 | 전국의 산지에 나는 여러해살이풀이다. 편평한 구형의 덩이줄기는 지름 2~4cm이고 주위에 작은 덩이줄기가 달려 번식한다. 윗부분에서 수염뿌리가 사방으로 퍼지고 얇은 막질의 인편이 겉을 감싸고 있다. 줄기는 곧추서고 녹색이며 백색 또는 자색 반점이 있다. 잎은 2장으로 5~14개의 소엽으로 갈라진다. 꽃은 5~6월에 피고 육수화서에 달린 불염포는 윗부분이 약간 자줏빛이 돌며 통부는 녹색이다. 육수화서의 연장부는 윗부분이 가는 원주형이다. 열매는 장과로 옥수수 모양이며 붉은색으로 익는다. |

분 포	한국, 일본, 만주, 우수리, 사할린
약재명	천남성(天南星)
성 분	triterphenesaponin, benzoic acid, 전분, amino acid
효 능	중풍, 반신불수, 구안와사, 수족마비, 전간, 파상풍, 자궁경부암

Description	Perennial herb, widely distributed in mountains. Tubers depressed, globose, 2-4cm in diameter, with small ones on the upper part, fibrous roots stretching, thin membranous scales enclosing the surface. Stems erect, green, dotted in white or purple. Leaves double, lobed into 5-14 leaflets. Flowers blooming in May-Jul. Spathes born in a spadix, slightly purple on the upper part, green on the tubular part; the extended part cylindrical, the top part attenuated. Fruits berry, corn-shaped, ripening in red.
Distribution	Korea, Japan, Manchuria, Ussuri, Sakhalin
Medicinal Name	Cheon-nam-seong(天南星)
Chemical components	Triterphenesaponin, benzoic acid, starch, amino acid
Effect	Stroke, hemiplegia, Bell's palsy, numbness in the hands and feet, epilepsy, tetanus, cervical cancer

이용부위 **Parts Used**
덩이줄기 Tuber

큰천남성

Arisaema ringens (Thunb.) Schott

천남성과 | Araceae

Puto Jackinthepulpit

| 식 물 | 남부지방의 산지에 나는 여러해살이풀이다. 편평한 구형의 덩이줄기는 지름 2~4cm이고 주위에 작은 덩이줄기가 달려 번식한다. 윗부분에서 수염뿌리가 사방으로 퍼지고 얇은 막질의 인편이 겉을 감싸고 있다. 줄기는 곧추서고 녹색이다. 잎은 2장으로 3개의 소엽으로 갈라진다. 표면은 윤채가 있는 녹색이며 뒷면은 흰빛을 띠고 끝이 실같이 가늘어진다. 꽃은 5월에 피며 육수화서에 달린 불염포는 접혀져서 등이 굽은 형상이고 겉은 녹색, 안쪽은 흑자색이다. 육수화서의 연장부는 곤봉 모양이다. 열매는 장과로 옥수수 모양이며 붉은색으로 익는다.

| 분 포 | 한국, 일본, 대만, 중국

| 약 재 명 | 천남성(天南星)

| 성 분 | triterphenesaponin, benzoic acid, 전분, amino acid

| 효 능 | 중풍, 반신불수, 구안와사, 수족마비, 전간, 파상풍, 자궁경부암

| Description | Perennial herb, widely distributed in mountains of southern peninsula. Tubers depressed, globose, 2-4cm in diameter, with small ones on the upper part, fibrous roots stretching, thin membranous scales enclosing the surface. Stems erect, green colored. Leaves double, lobed into 3 leaflets, lustrous green above, white-tinged beneath, attenuated at tips like thread. Flowers blooming in May-Jul. Spathes born in a spadix, folded, like the bent back, green outside, black purple inside; the extended part club-shaped. Fruits berry, corn-shaped, ripening in red.

| Distribution | Korea, Japan, Taiwan, China

| Medicinal Name | Cheon-nam-seong(天南星)

| Chemical components | Triterphenesaponin, benzoic acid, starch, amino acid

| Effect | Stroke, hemiplegia, Bell's palsy, numbness in the hands and feet, epilepsy, tetanus, cervical cancer

| 이용부위 Parts Used |
덩이줄기 Tuber

반 하
Pinellia ternata (Thunb.) Breitenb.
천남성과 | Araceae

식 물	전국의 밭이나 들판에 나는 여러해살이풀로 높이 30cm 정도이다. 땅속에 지름 1cm 가량의 덩이줄기가 있고 1~2개의 잎이 자라며 밑부분이나 위쪽에 1개의 주아가 생겼다가 떨어져서 번식한다. 잎은 3개의 소엽이 모여 달린다. 꽃은 6~7월에 육수화서로 달리고 암꽃은 밑에, 수꽃은 위에 달리며 육수화서의 연장부는 비스듬히 선다. 열매는 장과로 녹색이다.
분 포	한국, 일본, 대만, 만주, 중국
약 재 명	반하(半夏)
성 분	정유, 지방, 전분, 점액질
효 능	담, 가래, 해수, 천식

Description	Perennial herb, relatively rare in cultivated fields or grassy areas, 30cm tall. Tuber 1cm in diameter, bearing 1-2 leaves; gemma for propagation born on the bottom or upper petiole. Leaves three leaflets in cluster. Flowers born in a spadix, blooming in Jun.-Jul.; female flowers below, male flowers above. The extended part of spadix erect, slanting. Fruits berry, green.
Distribution	Korea, Japan, Taiwan, Manchuria, China
Medicinal Name	Ban-ha(半夏)
Chemical components	Essential oil, fat, starch, phlegmatic temperament
Effect	Sputum, cough, asthma

| 이용부위 Parts Used |
덩이줄기 Tuber

닭의장풀

Commelina communis L.
닭의장풀과 | Commelinaceae
Dayflower, Common Dayflower

식 물	전국의 길가나 풀밭, 냇가의 습지에 흔하게 자라는 한해살이풀로 가는 수염뿌리가 있다. 줄기의 아래쪽은 누워 자라고 마디에서 뿌리를 내린다. 흔히 가지가 갈라지며 높이 15~50cm 정도이다. 잎은 어긋나고 난상 피침형이다. 꽃은 7~8월에 하늘색으로 피고 엽액에서 나온 꽃줄기에 달린다. 열매는 삭과로 타원형이고 육질이며 익으면 3개로 갈라진다.
분 포	한국, 일본, 대만, 중국, 만주, 아무르, 우수리, 시베리아, 코카서스, 북미
약재명	압척초(鴨跖草)
성 분	chelidonine, protopine, stylopine, allocryptopine, berberine
효 능	해열, 이뇨작용, 당뇨병

Description	Annual herb, widely distributed on roadsides, grassy fields or wet places of streamside. Roots fibrous. Stems creeping, knotty, bearing roots, commonly branched, 15-50cm long. Leaves alternate, oval lanceolate. Flowers sky blue, born from the leaf axil of peduncle, blooming in Jul.-Aug. Fruits capsule, oblong, fleshy, cracked, opening in three valves after ripening.
Distribution	Korea, Japan, Taiwan, China, Manchuria, Amur, Ussuri, Siberia, Caucasus, North America
Medicinal Name	Ap-cheok-cho(鴨跖草)
Chemical components	Chelidonine, protopine, stylopine, allocryptopine, berberine
Effect	Antipyretic, diuretic effect, diabetes mellitus

이용부위 Parts Used
지상부 Above-ground parts

달래

Allium monanthum Maxim.
백합과 | Liliaceae
Uniflower Onion

식 물	전국의 산지 그늘진 곳에 모여 자라는 여러해살이풀이다. 비늘줄기는 넓은 난형으로 겉 비늘이 두껍고 밑에는 수염뿌리가 있다. 잎은 1~2개이며 선형 또는 넓은 선형이며 밑부분이 줄기를 둘러싸고 있다. 꽃은 4월에 흰색 또는 붉은빛이 도는 흰색으로 피고 줄기 끝에 1~2개가 달린다. 열매는 삭과로 작고 구형이다.
분 포	한국, 일본, 중국 동북부, 우수리강
약 재 명	소산(小蒜)
성 분	scorodose, allyl 계통의 화합물
효 능	토사곽란, 종기, 독충, 협심통

Description	Perennial herb, clumps in shady places of mountains. Bulbs broadly ovate; the surface is thick, with fibrous roots on the bottom. Leaves 1-2, linear or broadly linear, enclosing stems on the lower part. Flowers white or red-tinged white, born at the tip of stems with 1-2, blooming in Apr. Fruits capsule, small, globose.
Distribution	Korea, Japan, Northeast China, Ussuri River
Medicinal Name	So-san(小蒜)
Chemical components	Scorodose, allyl type compounds
Effect	Acute gastroenteritis, boil, venomous insect, anginal pain

| 이용부위 Parts Used |
비늘줄기 Bulb

두메부추 *Allium senescens* L.
백합과 | Liliaceae
Aging Onion

식 물	울릉도나 강원도의 해안가 절벽에 나는 여러해살이풀이다. 비늘줄기는 좁은 난형이고 섬유는 없으며 뿌리줄기가 길게 발달한다. 잎은 비늘줄기에서 4~9개가 나오며 선형으로 납작하고 육질이다. 줄기는 곧추서고 단면이 납작하며 날개가 있다. 꽃은 9~10월에 줄기 끝에서 피고 붉은빛을 띤 분홍색이다. 열매는 삭과, 종자는 검은색으로 타원형이다.
분 포	한국, 중국, 몽골, 러시아, 유럽
약재명	산구(山韭)
효 능	무기력증, 빈뇨증, 염증

Description	Perennial herb, rare on coastal cliffs of Ulleung Island or Gangwon-do. Bulbs narrowly ovate, fiberless, with long root stocks. Leaves 4-9, from a bulb, linear, flat, fleshy. Stems erect, compressed, with wings. Flowers red-tinged pink, born at the tip of stems, blooming in Sep.-Oct. Fruits capsule. Seeds black, oblong.
Distribution	Korea, China, Mongolia, Russia, Europe
Medicinal Name	San-gu(山韭)
Effect	Asthenia, micturition frequency, inflammation

| 이용부위 Parts Used |
비늘줄기, 지상부
Bulb and above-ground parts

산마늘

Allium microdictyon Prokh.
백합과 | Liliaceae
Victor Onion

식 물	깊은 산의 습한 곳에 드물게 자라는 여러해살이풀이다. 비늘줄기는 피침형으로 겉은 섬유로 둘러싸여 있다. 잎은 2~3개로 좁은 타원형 또는 타원형이며 잎의 기부는 줄기를 둘러싸고 있다. 꽃은 5~7월에 피며 노란빛을 띤 백색이다. 열매는 삭과, 종자는 검은색으로 구형이다.
분 포	한국, 중국, 몽골, 카자흐스탄, 시베리아
약재명	각총(茖葱)
성 분	methylallyldisulfide, diallyldisulfide, methylallyltrisulfide
효 능	감기

Description	Perennial herb, rare in wet places of deep mountains. Bulbs lanceolate, enclosed with fibers on the surface. Leaves 2-3, narrowly elliptical or elliptical; the basal part encloseing stems. Flowers yellow-tinged white, blooming in May-Jul. Fruits capsule. Seeds black, globose.
Distribution	Korea, China, Mongolia, Kazakhstan, Siberia
Medicinal Name	Gak-chong(茖葱)
Chemical components	Methylallyldisulfide, diallyldisulfide, methylallyltrisulfide
Effect	Cold

| 이용부위 **Parts Used** |
비늘줄기 Bulb

산부추

Allium thunbergii G.Don
백합과 | Liliaceae
Thunberg Onion

식 물	제주도를 제외한 전국 산지의 능선부 또는 양지바른 바위틈에 자라는 여러해살이풀이다. 줄기는 곧추서고 비늘줄기는 난형이며 검은빛을 띤 갈색의 섬유가 감싸고 있다. 잎은 3~5개가 비늘줄기에서 나오고 밑부분은 줄기를 감싸며 선형으로 단면은 대개 편평하다. 꽃은 9~10월에 줄기 끝에서 보라색 또는 짙은 보라색으로 핀다. 열매는 삭과, 종자는 검고 타원형이다.
분 포	한국, 일본, 대만, 중국
약재명	산구(山韭)
효 능	무기력증, 빈뇨증, 염증

Description	Perennial herb, relatively rare in rock crevices or in open sunny places of mountains in the peninsula. Stems erect. Bulbs ovate, enclosed with black-tinged brown fibers. Leaves 3-5 from bulbs, linear, mostly compressed; the basal part enclosing stems. Flowers purple or dark purple, born at the tip of stems, blooming in Sep.-Oct. Fruits capsule. Seeds black, oblong.
Distribution	Korea, Japan, Taiwan, China
Medicinal Name	San-gu(山韭)
Effect	Asthenia, micturition frequency, inflammation

| 이용부위 Parts Used |
비늘줄기, 지상부
Bulb and above-ground parts

울릉산마늘

Allium ochotense Prokh.
백합과 | Liliaceae

식 물	울릉도의 숲 속에 자라는 여러해살이풀이다. 비늘줄기는 피침형으로 겉은 섬유로 둘러싸여 있다. 잎은 2~3개로 타원형 또는 넓은 타원형이며 긴 잎집이 줄기를 둘러싸고 있다. 꽃은 5~7월에 피고 백색 또는 붉은빛이 도는 백색이다. 열매는 삭과, 종자는 검은색으로 구형이다.
분 포	한국, 중국, 사할린, 캄차카
약 재 명	각총(茖葱)
성 분	methylallyldisulfide, diallyldisulfide, methylallyltrisulfide
효 능	감기

Description	Perennial herb, common in the forest of Ulleung Island. Bulbs lanceolatee, enclosed with fibers on the surface. Leaves 2-3, elliptical or broadly elliptical; long leaf sheaths enclosing stems. Flowers white or red-tinged white, blooming in May-Jul. Fruits capsule. Seeds black, globose.
Distribution	Korea, China, Sakhalin, Kamchatka
Medicinal Name	Gak-chong(茖葱)
Chemical components	Methylallyldisulfide, diallyldisulfide, methylallyltrisulfide
Effect	Cold

| 이용부위 Parts Used |
비늘줄기 Bulb

참산부추 *Allium sacculiferum* Maxim.
백합과 | Liliaceae

이용부위 **Parts Used**
지상부, 비늘줄기
Above-ground parts and bulb

식 물	제주도를 제외한 전국 산지의 숲 가장자리나 기슭의 양지바른 곳에 자라는 여러해살이풀이다. 줄기는 곧추서며 비늘줄기는 난형이고 겉은 검은빛을 띤 갈색 섬유로 둘러싸여 있다. 잎은 3~5개가 비늘줄기에서 나오고 밑부분은 줄기를 감싸며 선형으로 단면은 각이 지거나 편평하다. 꽃은 9~10월에 줄기 끝에서 연보라색 또는 보라색으로 핀다. 열매는 삭과로 도심장형이고 종자는 검고 타원형이다.
분 포	한국, 일본, 중국, 러시아
약 재 명	산구(山韭)
효 능	무기력증, 빈뇨증, 염증

Description	Perennial herb, widely distributed (except the Jeju Island) in open sunny places in forest edges or the foothills of mountains. Stems erect. Bulbs ovate, enclosed with black-tinged brown fibers on the surface. Leaves 3-5 from bulbs, linear, angled or compressed the basal part enclosing stems. Flowers light purple or purple, born at the tip of stems, blooming in Sep.-Oct. Fruits capsule, obcordate. Seeds black, oblong.
Distribution	Korea, Japan, China, Russia
Medicinal Name	San-gu(山韭)
Effect	Asthenia, micturition frequency, inflammation

한라부추

Allium taquetii H. Lév.

백합과 | Liliaceae

| 식 물 | 제주도 한라산의 해발 1,000m 이상에서 자라는 여러해살이풀이다. 줄기는 곧추서며 비늘줄기는 난형으로 겉은 짙은 갈색의 섬유로 둘러싸여 있다. 3~5개의 잎이 비늘줄기에서 나오고 밑부분은 줄기를 감싸며 선형으로 단면은 원통형이나 드물게 각이 진다. 꽃은 9~10월에 줄기 끝에서 홍자색 또는 보라색으로 핀다. 열매는 삭과, 도심장형이고 종자는 검고 타원형이다.
| 분 포 | 특산(제주도)
| 약재명 | 구자(韭子-종자), 구채(韭菜-잎), 구근(韭根-비늘줄기)
| 성 분 | alkaloid, saponin
| 효 능 | 구자-양기쇠약증, 이뇨증
구채-관상 동맥 장애, 위장암, 토혈, 코피, 소변 출혈
구근-통증, 소화불량, 대하

| Description | Perennial herb, relatively common in 1,000m sea level of Mt. Halla in Jeju Island. Stems erect. Bulbs ovate, enclosed with dark brown fibers on the surface. Leaves 3-5 from bulbs, linear, cylindrical or infrequently angled the basal part enclosing stems. Flowers red purple or purple, born at the tip of stems, blooming in Sep.-Oct. Fruits capsule, obcordate. Seeds black, oblong.
| Distribution | Endemic (Jeju-do)
| Medicinal Name | Gu-ja(韭子 -seeds), gu-chae(韭菜 -leaf), gu-geun(韭根 -bulb)
| Chemical components | Alkaloid, saponin
| Effect | Guja - yang deficiency, enuresis; guchae - coronary artery disorder, stomach cancer, hemaetemesis, nosebleed, hematuria; gugeun - pain, dyspepsia, leukorrhea

| 이용부위 Parts Used |
종자, 잎, 뿌리(비늘줄기)
Seed, leaf, root and Bulb

윤판나물 *Disporum uniflorum* Baker
백합과 | Liliaceae

식 물	중부 이남의 숲 속에서 자라는 여러해살이풀로, 높이는 30~60cm 정도이다. 뿌리줄기는 옆으로 벋거나 짧고 줄기는 흔히 가지가 갈라진다. 잎은 어긋나게 달리고 3~5맥이 있다. 4~6월에 황색 꽃이 가지 끝에 1~3개가 달린다. 검정색으로 익는 열매는 장과로서 구형이다.
분 포	한국, 일본, 사할린
약 재 명	백미순(百尾笋)
효 능	기침, 천식, 가래, 복부팽만, 대장출혈, 장염, 폐결핵, 치질

Description	Perennial herb, common in forests south of the central peninsula, 30-60cm tall. Root stocks rhizomatous, short. Stems branched. Leaves alternate, with 3-5 veins. Flowers yellow, 1-3 flowers born at the tip of stems, blooming in Apr.-Jun. Fruits berry, globose, ripening in black.
Distribution	Korea, Japan, Sakhalin
Medicinal Name	Baeng-mi-sun(百尾笋)
Effect	Cough, asthma, sputum, abdominal fullness, enterohemorrhage, enteritis, tuberculosis, hemorrhoids

| 이용부위 Parts Used |
뿌리 Root

2004. SEUNG-HYUN YI.

큰애기나리 *Disporum viridescens* (Maxim.) Nakai
백합과 | Liliaceae
Virescent Fairybells

식 물	강원 이남 산지의 숲 속에서 자라는 여러해살이풀이다. 뿌리줄기는 옆으로 벋으면서 퍼진다. 줄기는 곧게 서며 가지가 갈라지고 높이는 30~70cm 정도이다. 잎은 어긋나고 3~5맥이 발달하며 잎자루는 없다. 연한 녹색 꽃은 4~5월에 가지 끝에서 1~3개가 밑을 향해 달린다. 열매는 장과로 둥글며 검은색이다.
분 포	한국, 일본, 중국
약재명	보주초(寶珠草)
효 능	기침, 천식, 소화불량, 복부창만, 대장출혈, 폐결핵, 장염, 치질

Description	Perennial herb, common in forests south of Gangwon-do. Root stocks rhizomatous. Stems erect, branched, 30-70cm tall. Leaves alternate, with 3-5 prominent veins, sessile. Flowers light green, born downward at the tip of branches with 1-3, blooming in Apr.-May. Fruits berry, round, black.
Distribution	Korea, Japan, China
Medicinal Name	Bo-ju-cho(寶珠草)
Effect	Cough, asthma, dyspepsia, abdominal fullness, enterohemorrhage, tuberculosis, enteritis, hemorrhoids

| 이용부위 Parts Used |
뿌리 Root

말나리

Lilium distichum Nakai ex Kamib.
백합과 | Liliaceae
Kochang Lily

| 식 물 | 산지의 높은 지대에서 약 80cm 정도까지 자라는 여러해살이풀이다. 비늘줄기에서 원줄기가 1개씩 나와 곧게 자란다. 줄기의 중간 부분에 달린 잎은 6~20개가 돌려나고 줄기 위쪽의 잎은 어긋나며 돌려난 잎보다 작다. 노란빛이 도는 붉은색 꽃은 6~7월에 피고 줄기 끝에서 1~10개가 옆을 향해 달리며 화피 안쪽에 짙은 갈색이 섞인 자줏빛 반점이 있다. 열매는 삭과로서 도란형이고 3개의 능선이 있으며 10월에 익는다.

| 분 포 | 한국, 중국 동북부, 흑룡강성, 사할린, 캄차카

| 약 재 명 | 동북백합(東北百合)

| 효 능 | 기침, 불면증, 부종

| Description | Perennial herb, widely distributed, but rarely seen in mountain forests. high altitude, almost up to 80cm in height. Main stems single, emerging from bulbs, erect. Leaves 6-20 whorled at the middle of stems, leaves of the upper part alternate, smaller than the whorled ones. Flowers yellow-tinged red, dotted in purple mixed with dark brown inside perianth, 1-10 flowers born at the tip of stems, blooming in Jun.-Jul. Fruits capsule, obovate, with three ridges, ripening in Oct.

| Distribution | Korea, Northeast China, Amur River, Sakhalin, Kamchatka

| Medicinal Name | Dong-buk-baek-hap(東北百合)

| Effect | Cough, insomnia, edema

| 이용부위 Parts Used |
비늘줄기 Bulb

섬말나리

Lilium hansonii Leichtlin ex Baker

백합과 | Liliaceae

Hanson Lily, Japanese Turks-cap Lily

식 물	울릉도 및 북부지방의 산지에 자라는 여러해살이풀이다. 비늘줄기는 약간 붉은색을 띠고 원줄기는 50~100cm 정도이다. 줄기 중간 부분에 달린 잎은 4~12개씩 돌려나며 2~4층을 이루고 위로 갈수록 작은 잎이 어긋나기로 달린다. 꽃은 6~7월에 원줄기 끝과 가지 끝에서 4~12개 정도가 아래쪽을 향해 달린다. 화피는 붉은 빛이 도는 황색, 안쪽에 검붉은색 반점이 있고 뒤로 말린다. 열매는 삭과로 둥글며 9월에 익는다.
분 포	한국, 만주, 아무르, 우수리
약재명	동북백합(東北百合)
효 능	기침, 불면증, 부종

Description	Perennial herb, relatively rare in the forest of Ulleung Island or northern part of the peninsula. Bulbs slightly red. Main stems 50-100cm tall. Leaves 4-12 at the middle of stems whorled in 2-4 tiers, the upper leaves smaller, alternate. Flowers born downward at the tip of main stems or branches, blooming in Jun.-Jul. Perianth red-tinged yellow, dotted in blackish red inside, retroflexed. Fruits capsule, round, ripening in Sep.
Distribution	Korea, Manchuria, Amur, Ussuri
Medicinal Name	Dong-buk-back-hap(東北百合)
Effect	Cough, insomnia, edema

| 이용부위 Parts Used |
비늘줄기 Bulb

참나리

Lilium lancifolium Thunb.
백합과 | Liliaceae
Tiger Lily

식 물	전국의 산과 들에 자라는 여러해살이풀이다. 비늘줄기는 흰색으로 원형이며 밑에서 뿌리가 나온다. 줄기는 높이 1~2m 정도이고 검은빛이 도는 자주색 점이 빽빽이 나 있으며 어릴 때는 흰색 거미줄 같은 털이 있다. 잎은 촘촘히 어긋나고 엽액에 짙은 갈색 주아가 달린다. 짙은 황적색 바탕에 흑자색 반점을 갖는 꽃은 7~8월에 원줄기와 가지 끝에서 4~20개가 밑을 향해 달린다. 열매를 잘 맺지 못하고 주로 주아가 땅에 떨어져 번식한다.
분 포	한국, 일본, 중국, 사할린
약 재 명	백합(百合)
성 분	전분, 단백질, 지방, 당, 칼륨
효 능	진해, 강장

Description	Perennial herb, widely distributed in mountains and fields. Bulbs white, round, bearing roots at the bottom. Stems 1-2m tall, densely dotted in black-tinged purple; when young, with white cobweb-like hairs. Leaves closely alternate, with brownish gemmas at the leaf axils. Flowers dotted in black-purple on a yellow-red background, nodding, 4-20 flowers at the axils of the tip of main stems or branches, blooming in Jul.-Aug. Fruits sterile, gemmas falling for propagation.
Distribution	Korea, Japan, China, Sakhalin
Medicinal Name	Baek-hap(百合)
Chemical components	Starch, protein, fat, sugar, potassium
Effect	Cough suppression, tonic effect

이용부위 **Parts Used**
비늘줄기 Bulb

삿갓나물

Paris verticillata M.Bieb.
백합과 | Liliaceae
Verticillate Paris

| 식 물 | 전국의 산지 숲 속에 자라는 여러해살이풀이다. 뿌리줄기는 옆으로 길게 벋고 마디마다 잔뿌리가 몇 개씩 나온다. 줄기는 곧추서며 높이 20~40cm 정도이다. 잎은 피침형 또는 좁고 긴 타원형으로 6~8장이 돌려난다. 꽃은 6~7월에 녹색으로 피고 줄기 끝에 1개가 달려 위를 향한다. 열매는 장과로 둥글며 자줏빛을 띤 검은색이다. |

분 포	한국, 일본, 만주, 아무르, 우수리, 사할린, 시베리아
약 재 명	조휴(蚤休)
성 분	pariphyllin, dioscin
효 능	종기, 외상출혈, 어혈성통증, 천식, 만성기관지염

Description	Perennial herb, widely distributed in mountains. Root stocks rhizomatous, knotty, fibrous. Stems erect, 20-40cm tall. Leaves lanceloate or narrow, long elliptical, 6-8 leaves whorled. Flowers green, solitary, born upward at the tip of stems, blooming in Jun.-Jul. Fruits berry, round, purple-tinged black.
Distribution	Korea, Japan, Manchuria, Amur, Ussuri, Sakhalin, Siberia
Medicinal Name	Jo-hyu(蚤休)
Chemical components	Pariphyllin, dioscin
Effect	Boil, open injury and bleeding, blood stasis pain, asthma, chronic bronchitis

| 이용부위 **Parts Used** |
뿌리줄기 Root stock

각시둥굴레

Polygonatum humile Fisher. ex Maxim.
백합과 | Liliaceae
Small Solomonseal

식 물	전국의 깊은 산이나 들의 숲 가장자리 풀밭에 자라는 여러해살이풀이다. 뿌리줄기는 가늘고 길게 옆으로 벋는다. 각진 줄기는 곧추서며 높이 15~30cm 정도이다. 잎은 어긋나고 엽병이 없다. 녹색 빛이 도는 긴 종 모양의 흰색 꽃은 5~6월에 줄기와 엽액에서 1~2개씩 달린다. 열매는 장과로 둥글며 짙은 하늘색이다.
분 포	한국, 일본, 중국 동북부, 쿠릴열도, 사할린, 시베리아
약 재 명	소옥죽(小玉竹)
효 능	구갈, 가슴답답증, 마른기침, 요통

Description	Perennial herb, widely distributed but relatively observed on the grassland and edge of forests or fields. Root stocks rhizomatous, attenuate, long. Stems angled, erect, 15-30cm tall. Leaves alternate, sessile. Flowers green-tinged white, long bell-shaped, 1-2 flowers born at the axil, blooming in May-Jun. Fruits berry, round, dark sky-blue.
Distribution	Korea, Japan, Northeast China, Kurils, Sakhalin, Siberia
Medicinal Name	So-ok-juk(小玉竹)
Effect	Mouth dry, choking, dry cough, lumbago

| 이용부위 Parts Used |
뿌리줄기 Root stock

왕둥굴레

Polygonatum robustum (Korsh.) Nakai
백합과 | Liliaceae

식 물	울릉도 산지의 숲 속에서 자라는 여러해살이풀이다. 뿌리줄기는 굵고 길게 옆으로 벋으며 마디가 약간 있고 수염뿌리가 많다. 줄기는 위쪽이 기울어지고 단면이 둥글며 높이 1m 정도까지 자란다. 잎은 어긋나고 좁은 타원형이다. 긴 종 모양의 꽃은 6~7월에 줄기의 중간 부분부터 1~3개씩 엽액에 달린다. 열매는 장과로 구형이며 검게 익는다.
분 포	한국, 일본, 만주
약재명	옥죽(玉竹)
성 분	Convallamarin, convallarin, quercitol, 전분, vitamin A
효 능	기침, 오한, 구갈

Description	Perennial herb, common in the forest of Ulleung Island mountains. Root stocks rhizomatous, lengthened, thick, a few knotty, many fibrous. Stems slanting on the upper, round, up to 1m tall. Leaves alternate, narrowly elliptical. Flowers long bell-shaped, 1-3 flowers axillary on the middle of a stem. Fruits berry, globose, ripening in black.
Distribution	Korea, Japan, Manchuria
Medicinal Name	Ok-juk(玉竹)
Chemical components	Convallamarin, convallarin, quercitol, starch, vitamin A
Effect	Cough, chill, mouth dry

| 이용부위 Parts Used |
뿌리줄기 Root stock

층층둥굴레 *Polygonatum stenophyllum* Maxim.
백합과 | Liliaceae

| 식 물 | 경기도 이북에서 자라는 여러해살이풀로 높이는 30~90cm 정도이다. 굵은 뿌리줄기는 옆으로 길게 자란다. 잎은 넓은 선형으로 2~6개가 돌려난다. 종 모양의 꽃은 6월에 연한 황색으로 피고 엽액에 돌려난다. 열매는 장과로 둥글며 9월에 검게 익는다. |

| 분 포 | 한국, 중국 동북부, 우수리 |

| 약재명 | 황정(黃精) |

| 성 분 | 점액질, 전분, 당, amino acid |

| 효 능 | 마른기침, 허약체질, 소갈증, 권태감, 무력감, 식욕감퇴, 맥박미약 |

| Description | Perennial herb, relatively rare in the sandy soil of river banks above north of Gyeonggi-do, 30-90cm tall. Root stocks rhizomatous, lengthened and thickened. Leaves broadly linear, 2-6 leaves whorled. Flowers blooming in Jun., light yellow, bell-shaped, born at the leaf axils. Fruits berry, globose, ripening in black in Sep. |

| Distribution | Korea, Northeast China, Ussuri |

| Medicinal Name | Hwang-jeong(黃精) |

| Chemical components | Phlegmatic temperament, starch, sugar, amino acid |

| Effect | Dry cough, asthenic diathesis, diabetes mellitus, fatigue, asthenia, appetite decreased, weak pulse |

| 이용부위 Parts Used |
| 뿌리줄기 Root stock |

금강애기나리 *Streptopus ovalis* (Ohwi) F.T.Wang & Y.C.Tang
백합과 | Liliaceae

| 식 물 | 깊은 산에서 자라는 여러해살이풀이다. 뿌리줄기는 옆으로 길게 벋고 줄기는 약 25~50cm 정도이다. 가지가 갈라지고 윗부분이 옆으로 처진다. 잎맥이 5~7개인 잎은 어긋나고 잎자루가 없으며 아래 부분이 심장형으로 줄기를 감싸고 있다. 꽃은 4~6월에 연한 황백색으로 피고 자주색 반점이 있으며 줄기 끝에서 1~3개가 산형으로 달린다. 열매는 장과로 둥글고 붉은색이다.
| 분 포 | 한국, 중국 동북부
| 약재명 | 보주초(寶珠草)
| 효 능 | 기침, 천식, 소화불량, 복부창만, 대장출혈, 폐결핵, 장염, 치질

Description Perennial herb, widely distributed in mountain ridges. Root stocks rhizomatous, long. Stems 2-50cm tall, branched, nodding on the upper part. Leaves alternate, with 5-7 veins, sessile, the base cordate, enclosing stems. Flowers yellowish white, dotted in purple, 1-3 flowers born in umbel at the tip of stems, blooming in Apr.-Jun. Fruits berry, round, red.

Distribution Korea, Northeast China

Medicinal Name Bo-ju-cho(寶珠草)

Effect Cough, asthma, dyspepsia, abdominal fullness, enterohemorrhage, tuberculosis, enteritis, hemorrhoids

| 이용부위 Parts Used |
뿌리 Root

연 영 초

Trillium kamtschaticum Pall. ex Pursh
백합과 ǀ Liliaceae
Trillium

|식 물| 깊은 산의 계곡부에 자라는 여러해살이풀이다. 짧고 굵은 뿌리줄기에서 두터운 수염뿌리가 모여난다. 줄기는 곧추서며 높이 20~30cm 정도이고 아래쪽에 갈색의 비늘잎이 있다. 잎은 줄기 끝에 3장이 돌려난다. 꽃은 5~6월에 백색으로 피고 꽃줄기 끝에 1개가 비스듬히 위를 향한다. 열매는 장과로 구형이며 8~9월에 흑자색으로 익는다.

|분 포| 한국, 일본, 만주, 아무르, 우수리, 사할린, 캄차카

|약 재 명| 우아칠(芋兒七)

|성 분| trillin, trillarin, diosgenin, cyasterone, ecdysterone

|효 능| 두통, 어지럼증, 신경성두통, 요통, 하지신경통, 타박상, 골절상, 외상출혈

|Description| Perennial herb, relatively rare in the valleys of deep mountains above the middle of peninsula. Root stocks short, bulky root stocks, thick, in clusters, roots fibrous. Stems erect, 20-30cm tall; scaly leaves on the lower part brownish. Leaves whorled with three. Flowers white, solitary, born upward at the tip of peduncle, blooming in May.-Jun. Fruits berry, globose, black purple, ripening in Aug.-Sep.

|Distribution| Korea, Japan, Manchuria, Amur, Ussuri, Sakhalin, Kamchatka

|Medicinal Name| U-a-chi(芋兒七)

|Chemical components| Trillin, trillarin, diosgenin, cyasterone, ecdysterone

|Effect| Headache, dizziness, tension headache, lumbago, neuralgia, bruise, fracture, open injury and bleeding

|이용부위 **Parts Used** |
뿌리줄기 Root stock

산자고

Tulipa edulis (Miq.) Baker
백합과 | Liliaceae
Edible Tulip

|식 물| 중부 이남 산지의 양지바른 풀밭에 나는 여러해살이풀이다. 비늘줄기는 난상 원형이고 옅은 자갈색이며 밑에 수염뿌리가 있다. 잎은 비늘줄기에서 2장이 나고 선형으로 흰빛이 도는 녹색이다. 꽃은 4~5월에 피고 비늘줄기에서 나온 꽃줄기 끝에 1~3개가 달리며 흰색 바탕에 자줏빛 맥이 있다. 열매는 삭과로 녹색이며 거의 둥글고 세모가 진다.

|분 포| 한국, 일본, 만주, 중국

|약 재 명| 광자고(光慈姑)

|성 분| 점액질, glucomannan, 포도당

|효 능| 종기, 종창, 악창, 결핵성림프선염

|Description| Perennial herb, relatively common in open sunny grassland south of the central peninsula. Bulbs ovate, light purplish brown, with fibrous roots. Leaves double from bulbs, linear, white-tinged green. Flowers white, with violet veins, 1-3 flowers born at the tip of peduncle, blooming in Apr.-May. Fruits capsule, green, mostly round or triangular.

|Distribution| Korea, Japan, Manchuria, China

|Medicinal Name| Gwang-ja-go(光慈姑)

|Chemical components| Phlegmatic temperament, glucomannan, glucose

|Effect| Boil, swelling, obstinate abscess, tuberculous lymphadenitis

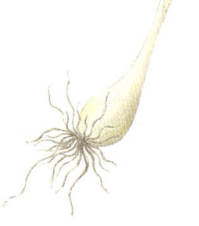

|이용부위 Parts Used|
비늘줄기 Bulb

자 란

Bletilla striata (Thunb.) Rchb.f.
난초과 | Orchidaceae
Common Bletilla

식 물	전라남도 산지의 양지바른 곳에 드물게 자라는 여러해살이풀이다. 덩이뿌리는 육질이며 옆으로 줄지어 붙는다. 잎은 덩이뿌리에서 나오고 밑부분에서 5~6개가 서로 감싸면서 줄기처럼 되며 긴 타원형이다. 꽃은 5~6월에 피며 잎 사이에서 나온 꽃줄기의 끝에 3~7개의 홍자색 꽃이 총상으로 달린다. 열매는 긴 타원형이다.
분 포	한국, 일본, 중국
약재명	백급(白芨)
성 분	수분, 전분, 포도당, 정유, 점액질
효 능	각혈, 토혈, 위·십이지장궤양, 외상, 종기, 창양

Description	Perennial herb, infrequent in the open sunny places of mountains in Jeollanam-do. Tubers fleshy, attached to each other in rows. Leaves born from tubers, long-elliptical; on the lower part, leaves 5-6, enclosing each other like a stem. Flowers red purple, born in raceme with 3-7, at the tip of peduncle from between leaves, blooming in May-Jun. Fruits long-oblong.
Distribution	Korea, Japan, China
Medicinal Name	Baek-geup(白芨)
Chemical components	Water, starch, glucose, essential oil, phlegmatic temperament
Effect	Hemoptysis, hematemesis, peptic ulcer, wound, boil, acute bacterial infection

| 이용부위 Parts Used |
덩이뿌리 Tuber

약난초

Cremastra variabilis (Blume) Nakai
난초과 | Orchidaceae
Appendiculate Cremastra

식 물	전라도 이남의 산지 숲 속에 드물게 나는 여러해살이풀이다. 위인경은 난상 원형이고 땅 속에 얕게 묻히며 옆으로 염주같이 연결된다. 잎은 위인경의 끝에서 1개(드물게 2개)가 나오며 긴 타원형이다. 꽃줄기는 위인경 옆에서 나오며 곧추서고 높이 30~50cm 정도이다. 꽃은 5~6월에 연한 자줏빛이 도는 갈색으로 피고 한쪽으로 치우쳐서 밑을 향해 달린다. 열매는 삭과로 긴 타원형이다.
분 포	한국, 일본, 대만, 중국, 히말라야
약 재 명	산자고(山慈姑)
성 분	점액질, 포도당, glucomannan
효 능	종기, 종창, 악창, 결핵성림프선염

Description	Perennial herb, infrequent in the forests of Jeolla-do. Pseudobulbs ovate, slightly covered under the ground, attached to each other like beads. Leaves long elliptical; one leaf (or infrequently two leaves) from pseudobulbs. Pedicels born from the side of pseudobulbs, erect, 30-50cm in height. Flowers light purple brown, nodding to one side, blooming in May-Jun. Fruits capsule, oblong.
Distribution	Korea, Japan, Taiwan, China, Himalaya
Medicinal Name	San-ja-go(山慈姑)
Chemical components	Phlegmatic temperament, glucose, glucomannan
Effect	Boil, swelling, obstinate abscess, tuberculous lymphadenitis

| 이용부위 Parts Used |
위인경 Pseudobulb

석곡

Dendrobium moniliforme (L.) Sw.
난초과 | Orchidaceae
Moniliforme Dendro-bium

식 물	남부지방과 도서지역의 바위나 죽은 나무줄기에 부착해 자라는 여러해살이품이다. 흰색의 굵은 뿌리가 많다. 줄기는 근경에서 여러 개가 뭉쳐나고 곧게 서며 다육질로 약 10~20cm 정도이다. 오래된 줄기는 잎이 없고 마디만 있으며 녹색을 띤 갈색이다. 짙은 녹색 잎은 어긋나고 끝이 둔하다. 꽃은 5~6월에 흰색이나 연한 적색으로 핀다. 열매는 삭과로 난형이다.
분 포	한국, 일본, 중국
약재명	석곡(石斛)
성 분	dendrobine, dendramine, nobilonine, dendroxine, dendrin
효 능	해열, 진통작용, 소화불량, 백내장, 요통

Description	Perennial herb, relatively rare on the shallow soils in rocks or dead trees in the southern area or islands. Roots white, thick, numerous. Stems in clusters, from a rhizome, erect, fleshy, almost 10-20cm tall; aged stems without leaves, knotty, greenish brown. Leaves dark green, alternate, obtuse on the tip. Flowers white or light red, blooming in May-Jun. Fruits capsule, ovate.
Distribution	Korea, Japan, China
Medicinal Name	Seok-gok(石斛)
Chemical components	Dendrobine, dendramine, nobilonine, dendroxine, dendrin
Effect	Antipyretic, pain relief, dyspepsia, cataract, lumbago

| 이용부위 Parts Used |
지상부 Above-ground parts

천 마

Gastrodia elata Blume
난초과 | Orchidaceae
Tall Gastrodia

식 물	거의 전국의 산지에 드물게 나는 여러해살이 부생식물이다. 덩이줄기는 비대하며 긴 타원형으로 옆으로 벋고 환상의 불분명한 선이 많다. 줄기는 높이 60~100cm 정도로 곧추서며 털이 없고 황갈색이다. 잎은 없다. 꽃은 6~7월에 피고 황갈색이며 많은 꽃이 모여 총상화서를 이룬다. 열매는 삭과로 도란형이다.
분 포	한국, 일본, 대만, 중국, 만주, 아무르, 우수리
약재명	천마(天麻)-덩이뿌리, 적전(赤箭)-싹
성 분	vanillin, vanillylalcohol, vitamin A, 점액질
효 능	천마-두통, 신경쇠약, 파상풍, 진정·항경련, 진통·항염증, 혈압강하, 항산화, 면역강화 적전-기력증진, 종기

| 이용부위 Parts Used |
| 덩이줄기, 싹(꽃줄기) |
| Tuber, flower bud, and flower stalk |

Description Perennial herb, widely distributed but rarely observed in mountains, a saprophyte. Tubers fleshy, oblong, with obscure annular lines. Stems 60-100cm tall, erect, glabrous, yellowish brown. Leaves absent. Flowers yellowish brown, many flowers born in a raceme, blooming in Jun.-Jul. Fruits capsule, obovate.

Distribution Korea, Japan, Taiwan, China, Manchuria, Amur, Ussuri

Medicinal Name Cheon-ma(天麻) - tuber, jeok-jeon(赤箭) - bud

Chemical components Vanillin, vanillyl alcohol, vitamin A, phlegmatic temperament

Effect Cheonma - headache, neurasthenia, tetanus, sedation and anticonvulsion, pain relief and anti-inflammation, blood pressure lowering, antioxidation, immunity enhancement; jeokjeon - energization, boil

용어해설 Glossary

유형	용어	해설	해당식물
식물체의 특성	관목	키가 2~3m 내외의 목본(木本) 식물로서 주간(主幹)이 분명하지 않고 밑동에서 가지가 많이 나는 나무. 진달래, 사철나무, 앵두나무 따위.	겨우살이, 가시오갈피, 누리장나무, 백리향, 섬백리향
	부생식물	어떤 생물체가 죽거나 썩은 뒤 그로부터 양분을 섭취하는 식물.	천마
	아교목	교목과 관목의 중간 식물. 교목보다 작지만 모양은 교목과 같음.	산수유
지하부	덩이줄기	괴경(塊莖, tuber): 덩이 모양을 이룬 땅속줄기의 한 가지. 영양물의 저장소로 되어 있어서 몹시 비대함.	점현호색, 현호색, 흑삼릉, 둥근잎천남성, 천남성, 두루미천남성, 점박이천남성, 큰천남성, 반하, 천마
	비늘줄기	인경(鱗莖, bulb): 땅속줄기의 한 가지. 줄기 자체가 비대하여 덩어리 또는 둥근 모양의 줄기가 된 것과는 다르게 줄기의 끝에 육질이 풍부한 다수의 마디가 짧은 줄기를 둘러싼 지하 저장기관.	산마늘, 달래, 울릉산마늘, 참산부추, 두메부추, 한라부추, 산부추, 말나리, 섬말나리, 참나리, 산자고, 약난초
	뿌리줄기	근경(根莖, root stock): 줄기가 변태된 지하경의 하나. 뿌리 비슷하게 땅 속으로 뻗어 나가며 많은 마디가 생기고 각 마디에서 부정근이 남.	고란초, 산일엽초, 우단일엽, 삼백초, 꿩의바람꽃, 삼지구엽초, 깽깽이풀, 큰괭이밥, 제비꽃, 참졸쌀풀, 용담, 과남풀, 선갈퀴, 산국, 흑삼릉, 두메부추, 윤판나물, 큰애기나리, 삿갓나물, 각시둥굴레, 왕둥굴레, 층층둥굴레, 금강애기나리, 연영초
	위인경	(僞鱗莖, pseudobulb): 난초과 식물의 줄기가 불룩해져서 인경(비늘줄기)처럼 보이는 것.	약난초
	인편	(鱗片, scale): 식물체의 겉면을 덮고 있는 비늘 모양의 조각.	우단일엽, 큰괭이밥, 둥근잎천남성, 천남성, 점박이천남성, 큰천남성
줄기의 특성	엽액	식물의 가지나 줄기와 잎자루의 사이. 잎겨드랑이.	쥐방울덩굴, 둥칡, 으름덩굴, 뱀딸기, 다래, 참졸쌀풀, 용담, 과남풀, 박주가리, 벌깨덩굴, 백리향, 섬백리향, 인동, 흑삼릉, 닭의장풀, 참나리, 각시둥굴레, 왕둥굴레, 층층둥굴레
	포복성	식물의 줄기가 기고, 뻗고, 퍼지며, 휘감기는 성질.	좀씀바귀, 뱀딸기
잎의 특성	근생엽	뿌리에서나 지하경에서 나온 잎.	큰엉겅퀴

유형	용어	해설	해당식물
잎의 특성	비늘잎	비늘 같은 잎으로 줄기의 밑동을 쌈.	연영초
	선점	생물체 내에서 분비작용을 하는 기관.	백리향, 섬백리향
	소엽	겹잎을 이루는 작은 잎.	꿩의바람꽃, 으름덩굴, 뱀딸기, 둥근잎천남성, 천남성, 두루미천남성, 점박이천남성, 큰천남성, 반하
	잎집	잎의 하단부에서 줄기를 둘러싸고 있는 부분. 벼, 보리, 옥수수와 같은 화본과(禾本科) 식물에 많음. 잎싸개.	울릉산마늘
	장상엽	손바닥을 편 모양으로 발달된 복엽.	으름덩굴, 가시오갈피, 인삼
	주맥	잎의 한 가운데 있는 가장 큰 잎맥.	우단일엽, 선갈퀴
잎이 달리는 모양	돌려나기	윤생(輪生, whorled): 한 마디에 세 개 이상의 잎이 돌아가며 붙은 상태.	참좁쌀풀, 선갈퀴, 도라지, 말나리, 섬말나리, 산수유
	마주나기	대생(對生, opposite): 한 마디에 한 쌍의 잎이 서로 마주 보면서 달린 모양.	패랭이꽃, 참좁쌀풀, 용담, 과남풀, 큰구슬붕이, 박주가리, 누리장나무, 꽃향유, 벌깨덩굴, 백리향, 섬백리향, 인동, 도라지, 산국, 흑삼릉
	어긋나기	호생(互生, alternate): 잎이 줄기의 마디마다 하나씩 붙은 상태.	삼백초, 쥐방울덩굴, 으름덩굴, 애기똥풀, 점현호색, 현호색, 뱀딸기, 다래, 가시오갈피, 큰까치수염, 도라지, 벌개미취, 삽주, 큰엉겅퀴, 바늘엉겅퀴, 좀씀바귀, 왕씀배, 닭의장풀, 윤판나물, 큰애기나리, 말나리, 섬말나리, 참나리, 각시둥굴레, 왕둥굴레, 금강애기나리, 석곡
잎의 모양	난형	(卵形, ovate): 달걀 모양.	삼백초, 으름덩굴, 삼지구엽초, 뱀딸기, 섬시호, 참좁쌀풀, 큰구슬붕이, 누리장나무, 꽃향유, 도라지, 좀씀바귀, 미역취
	선형	(線形, linear): 선처럼 가늘고 긴 모양.	우단일엽, 벌개미취, 흑삼릉, 달래, 참산부추, 두메부추, 한라부추, 산부추, 층층둥굴레, 산자고
	심장형	(心臟形, cordate): 식물의 잎이 심장 모양.	쥐방울덩굴, 등칡, 삼지구엽초, 박주가리, 벌깨덩굴, 금강애기나리

용어해설 Glossary

유형	용어	해설	해당식물
잎의 모양	타원형	(楕圓形, elliptical): 길이가 폭의 두 배 정도 되는 길고 둥근 모양.	으름덩굴, 다래, 백리향, 섬백리향, 선갈퀴, 인동, 좀쐐기, 미역취, 산마늘, 울릉산마늘, 삿갓나물, 왕둥굴레, 자란, 약난초
	피침형	(披針形, lanceolate): 창처럼 생긴 모양. 길이는 너비의 몇 배가 되고 밑에서 1/3정도 되는 부분이 가장 넓으며 끝이 뾰족한 것.	고란초, 산일엽초, 제비꽃, 인동, 도라지, 벌개미취, 미역취, 닭의장풀, 삿갓나물
화서의 유형	두상화	(頭狀花, capitulum): 꽃대 끝에 많은 꽃이 붙어 머리 모양을 이룬 꽃.	삽주, 큰엉겅퀴, 미역취
	산방화서	(繖房花序, corymb): 위쪽에 난 꽃자루보다 아래쪽에 난 꽃자루가 길기 때문에 꽃차례를 이루는 꽃들이 전체적으로 거의 평면으로 배열하게 되는 것.	
	산형화서	(傘形花序, umbel): 꽃차례 축의 한 지점에서 작은 꽃자루를 갖는 꽃들이 방사상으로 배열한 꽃차례. 편평하거나 구형을 이룸.	섬시호, 산수유, 두메부추, 산부추, 한라부추
	육수화서	(肉穗花序, spadix): 주축이 육질이고 꽃자루가 없는 많은 수의 작은 꽃이 밀집한 꽃차례.	둥근잎천남성, 천남성, 두루미천남성, 점박이천남성, 큰천남성, 반하
	총상화서	(總狀花序, raceme): 긴 꽃차례 축에 꽃자루가 있는 여러 개의 꽃이 어긋나게 붙어 있는 꽃차례.	삼백초, 으름덩굴, 삼지구엽초, 점현호색, 현호색, 큰까치수염, 박주가리, 자란, 천마
	취산화서	(聚散花序, cyme): 화서의 중앙에 있는 꽃이 먼저 핀 다음, 뒤이어 그 옆에 있는 꽃이 피는 것.	선갈퀴
꽃의 특성	불염포	(佛焰苞, spathe): 넓은 잎과 같은 모양의 포로 육수화서를 둘러싸고 있는 구조.	둥근잎천남성, 천남성, 두루미천남성, 점박이천남성, 큰천남성, 반하
	포자낭군	(胞子囊群, sorus): 양치류의 잎 뒷면에 모여 있는 포자낭의 무리.	고란초, 산일엽초, 우단일엽
	화관	(花冠, corolla): 꽃받침 안쪽에 있는 꽃잎을 총칭하는 말 = 꽃부리.	선갈퀴
	화피	(花被, perianth): 꽃받침과 꽃잎의 구별이 명확하지 않은 꽃에서 이들을 총칭하는 말.	말나리, 섬말나리

유형	용어	해설	해당식물
열매의 종류	견과	(堅果, nut): 도토리나 호두처럼 껍질이 단단한 열매. 크기가 작은 견과를 소견과(nutlet)라고도 함.	꽃향유, 벌깨덩굴, 백리향, 섬백리향
	골돌	(骨突, follicle): 열매가 다 익게 되면 배봉선을 따라 벌어짐. 한 꽃에 여러 개의 암술이 익어서 하나의 열매가 되기도 함.	삼지구엽초, 박주가리
	분열과	(分裂果, schizocarp): 중축(中軸, stylopodium)에 2~여러 개의 분과(分果, mericarp)가 달려 있다가 성숙하면 각각 떨어져 나감.	섬시호
	삭과	(蒴果, capsule): 합생심피 자방이 성숙해서 속이 여러 칸으로 나뉘고 각 칸에 많은 종자가 들어 있는 열매.	삼백초, 쥐방울덩굴, 등칡, 패랭이꽃, 깽깽이풀, 애기똥풀, 점현호색, 현호색, 큰괭이밥, 제비꽃, 큰까치수염, 참좁쌀풀, 용담, 과남풀, 큰구슬붕이, 도라지, 닭의장풀, 산마늘, 달래, 울릉산마늘, 참산부추, 두메부추, 한라부추, 산부추, 말나리, 섬말나리, 참나리, 산자고, 약난초, 석곡, 천마
	수과	(瘦果, achen): 1개의 종자가 들어 있으며 날개가 있거나 깃털이 있고 익어도 벌어지지 않는 열매.	꿩의바람꽃, 뱀딸기, 벌개미취, 삽주, 큰엉겅퀴, 바늘엉겅퀴, 산국, 왕씀배, 미역취
	장과	(漿果, berry): 종자가 육질의 과피 속에 매몰되어 있는 열매.	겨우살이, 다래, 인동, 둥근잎천남성, 천남성, 두루미천남성, 점박이천남성, 큰천남성, 반하, 윤판나물, 큰애기나리, 삿갓나물, 각시둥굴레, 왕둥굴레, 금강애기나리, 연영초
	핵과	(核果, drupe): 다육질의 껍질이 있고 하나의 종자가 단단한 내과피로 둘러싸여 있는 열매.	산수유, 누리장나무
	과피	(果皮, pericarp): 열매의 껍질 부분.	으름덩굴
	관모	(冠毛, pappus): 수과의 머리에 붙어 있는 털뭉치. 국화과에서는 꽃받침이 털로 변한 것.	삽주, 큰엉겅퀴, 바늘엉겅퀴, 좀씀바귀, 왕씀배
	주아	(珠芽, gemma): 씨앗이나 열매가 아니면서 식물체의 몸 일정 부위에 생겨나 나중에 새로운 하나의 식물로 자라나는 조직 = 살눈.	반하, 참나리

찾아보기 Index

한글명 찾아보기 Korean

ㄱ
가시오갈피 54
각시둥굴레 148
겨우살이 24
고란초 16
과남풀 68
금강애기나리 154
깽깽이풀 38
꽃향유 76
꿩의바람꽃 32

ㄴ
누리장나무 74

ㄷ
다래 50
달래 122
닭의장풀 120
도라지 88
두루미천남성 112

두메부추 124
둥근잎천남성 108
등칡 26

ㅁ
말나리 140
미역취 104

ㅂ
바늘엉겅퀴 96
박주가리 72
반하 118
백리향 80
뱀딸기 46
벌개미취 90
벌깨덩굴 78

ㅅ
산국 98

산마늘 126
산부추 128
산수유 60
산일엽초 18
산자고 158
삼백초 22
삼지구엽초 36
삽주 92
삿갓나물 146
석곡 164
선갈퀴 84
섬말나리 142
섬백리향 82
섬시호 58

ㅇ
애기똥풀 40
약난초 162
연영초 156
왕둥굴레 150
왕씀배 102

용담 66
우단일엽 20
울릉산마늘 130
윤판나물 136
으름덩굴 34
인동 86
인삼 56

ㅈ
자란 160
점박이천남성 114
점현호색 42
제비꽃 52
좀씀바귀 100
쥐방울덩굴 28

ㅊ
참나리 144
참산부추 132
참좁쌀풀 62

천남성 110
천마 166
층층둥굴레 152

ㅋ
큰괭이밥 48
큰구슬붕이 70
큰까치수염 64
큰애기나리 138
큰엉겅퀴 94
큰천남성 116

ㅍ
패랭이꽃 30

ㅎ
한라부추 134
현호색 44
흑삼릉 106

학명 찾아보기 Scientific name

A
Actinidia arguta 50
Akebia quinata 34
Allium microdictyon 126
Allium monanthum 122
Allium ochotense 130
Allium sacculiferum 132

Allium senescens 124
Allium taquetii 134
Allium thunbergii 128
Anemone raddeana 32
Arisaema amurense 108
Arisaema amurense for. serratum 110
Arisaema heterophyllum 112

Arisaema peninsulae 114
Arisaema ringens 116
Aristolochia contorta 28
Aristolochia manshuriensis 26
Asperula odorata 84
Aster koraiensis 90
Atractylodes ovata 92

B

Bletilla striata 160
Bupleurum latissimum 58

C

Chelidonium majus var.
  asiaticum 40
Cirsium pendulum 94
Cirsium rhinoceros 96
Clerodendrum trichotomum 74
Commelina communis 120
Cornus officinalis 60
Corydalis maculata 42
Corydalis remota 44
Cremastra variabilis 162
Crypsinus hastatus 16

D

Dendranthema boreale 98
Dendrobium moniliforme 164
Dianthus chinensis 30
Disporum uniflorum 136
Disporum viridescens 138
Duchesnea indica 46

E

Eleutherococcus senticosus 54
Elsholtzia splendens 76
Epimedium koreanum 36

G

Gastrodia elata 166
Gentiana scabra 66
Gentiana triflora var. japonica 68
Gentiana zollingeri 70

I

Ixeris stolonifera 100

J

Jeffersonia dubia 38

L

Lepisorus ussuriensis 18
Lilium distichum 140
Lilium hansonii 142
Lilium lancifolium 144
Lonicera japonica 86
Lysimachia clethroides 64
Lysimachia coreana 62

M

Meehania urticifolia 78
Metaplexis japonica 72

O

Oxalis obtriangulata 48

P

Panax ginseng 56
Paris verticillata 146
Pinellia ternata 118
Platycodon grandiflorum 88
Polygonatum humile 148
Polygonatum robustum 150
Polygonatum stenophyllum 152
Prenanthes ochroleuca 102
Pyrrosia linearifolia 20

S

Saururus chinensis 22
Solidago virgaurea subsp.
  asiatica 104
Sparganium erectum 106
Streptopus ovalis 154

T

Thymus quinquecostatus 80
Thymus quinquecostatus var.
  japonica 82
Trillium kamtschaticum 156
Tulipa edulis 158

V

Viola mandshurica 52
Viscum album var. coloratum 24

찾아보기 Index

영문명 찾아보기 Common name

A
Aging Onion 124
Appendiculate Cremastra 162
Asian Celandine 40
Asian Greater Celandine 40

B
Balloon Flower 88
Bower Actinidia 50
Bur Reed 106

C
Chinese Bellflower 88
Chinese Lizard's Tail 22
Chinese Pink 30
Chinese Twinleaf 38
Chocolate Vine 34
Common Bletilla 160
Common Dayflower 120
Corydalis 44

D
Dagalet Thyme 82
Dayflower 120
Diversileaf Jackinthepulpit 112

E
Edible Tulip 158

F
Five-Leaf Akebia 34
Fiveribbed Thyme 80

G
Gentian 66
Ginseng 56
Glory Bower 74
Gold-and-silver Flower 86
Goldenrod 104
Gooseneck Loosestrife 64

H
Haichow Elsholtzia 76
Hanson Lily 142
Harlequin Glory-bower 74

J
Japanese Atractylodes 92
Japanese Cornel 60
Japanese Cornelian Cherry 60
Japanese Honeysuckle 86
Japanese Metaplexis 72
Japanese Turks-cap Lily 142

K
Knope-sedge 106
Kochang Lily 140
Korean Daisy 90
Korean Epimedium 36
Korean Jackinthepulpit 114
Korean Loosestrife 62
Korean Starwort 90

M
Manchurian Birthwort 26
Manchurian Dutchman's pipe 26
Manchurian Violet 52
Mistletoe 24
Moniliforme Dendrobium 164
Mountain Lady's-sorrel 48

N
Nettleleaf Meehania 78
North Chrysanthemum 98
Northern Dutchmanspipe 28

O
Officinalis Dogwood 60

P
Pendulate Thistle 94
Pink 30
Puto Jackinthepulpit 116

R
Radde Anemone 32

S
Serrate Amur Jackinthepulpit 110
Siberian ginseng 54
Small Solomonseal 148

T
Tall Gastrodia 166
Tara Vine 50
Threeflower Gentian 68
Thunberg Onion 128
Thyme 80
Tiger Lily 144
Trillium 156

U
Uniflower Onion 122

V
Verticillate Paris 146
Victor Onion 126
Vine Pear 50
Virescent Fairybells 138

W
Woodruff 84

Y
Yang-tao 50

Z
Zollinger Gentian 70

참고문헌 References

강병수, 이장천, 주영승, 오수석, 박용기. 2008. 『원색 한약도감』. 동아문화사.
강병화. 2008. 『한국생약자원생태도감 1, 2, 3』. 지오북.
국립수목원. 1997. 『희귀 및 멸종위기 식물도감』. 생명의 나무.
국립수목원. 2005. 『세밀화로 보는 광릉숲의 풀과 나무』. 김영사.
국립수목원, 한국식물분류학회. 2007. 『국가표준식물목록』. 대신기획.
문순화, 현진오. 2003. 『봄에 피는 우리꽃 386』. 신구문화사.
문순화, 현진오. 2003. 『여름에 피는 우리꽃 386』. 신구문화사.
문순화, 현진오. 2004. 『가을에 피는 우리꽃 336』. 신구문화사.
신현철, 최홍근. 1997. '한국산 백리향속 식물의 분류학적 연구: 수리분류학적 접근'. 『한국식물분류학회지』. 27(2): 117~135.
안덕균. 1998. 『원색 한국본초도감』. 교학사.
오병운. 1999. '한국산 현호색속(*Corydalis*, Fumariaceae)의 분류학적 재검토'. 『한국식물분류학회지』. 29(3): 201~230.
오병운, 조동광, 김규식, 장창기. 2005. 『한반도 특산 관속식물』. 국립수목원.
이영노. 2006. 『새로운 한국식물도감 I, II』. 교학사.
이우철. 1996. 『원색한국기준식물도감』. 아카데미서적.
이우철, 임양재. 2002. 『식물지리』. 강원대학교 출판부.
이유성. 1999. 『현대식물분류학』. 우성문화사.
이창복. 2003. 『원색 대한식물도감 상, 하』. 향문사.
장창기. 2002. '한국산 둥굴레속(*Polygonatum*, Ruscaceae)의 분류학적 재검토'. 『한국식물분류학회지』. 32(4): 417~447.
최혁재, 장창기, 고성철, 오병운. 2004. '한국산 부추속(*Allium*, Alliaceae)의 분류학적 재검토'. 『한국식물분류학회지』. 34(2): 119~152.
한국양치식물연구회. 2005. 『한국양치식물도감』. 지오북.
Ko, S.C., Y.S. Kim. 1985. A taxonomic study on genus *Arisaema* in Korea. Kor. J. Plant Tax. 15(2): 67~109.

http://www.encyber.com/-두산백과사전
http://www.kna.go.kr:9320/-한국의 희귀식물
http://www.koreaplants.go.kr:9101/-국가표준식물목록
http://www.nature.go.kr/-국가생물종지식정보시스템

세밀화로 보는 약용식물
Botanical Art of Korean Medicinal Plants

초판 1쇄 발행 2010년 1월 30일
초판 4쇄 발행 2017년 8월 25일

지은이 　국립수목원

집필자 　이정희, 장창석, 이유미, 조동광
세밀화 　권순남, 공혜진, 이승현, 강영인, 허설희, 서지연, 구순원, 김경숙, 정인영

펴낸곳 　지오북(GEOBOOK)
펴낸이 　황영심
디자인 　김길례

주소 　서울특별시 종로구 사직로8길 34, 오피스텔 1018호
　　　(내수동 경희궁의아침 3단지)
　　　Tel_ 02-732-0337
　　　Fax_ 02-732-9337
　　　eMail_ geo@geobook.co.kr
　　　www.geobook.co.kr

출판등록번호 　제300-2003-211
출판등록일 　2003년 11월 27일

ⓒ 국립수목원, 지오북 2010
지은이와 협의하여 검인은 생략합니다.

ISBN 978-89-94242-00-2 03600

*이 책은 저작권법에 따라 보호받는 저작물입니다. 이 책 내용과 사진의 저작권에 대한 문의는 지오북(GEOBOOK)으로 해주십시오.

**이 책은 우리나라에서 자생하는 약용식물을 소개하는 책이므로 약재의 처방과 이용은 전문가의 지시에 따르시기 바랍니다.